Soil Fertility Management

Soil Fertility Management

Mr. Guru Sharan Das

RANDOM PUBLICATIONS
NEW DELHI (INDIA)

Soil Fertility Management

ISBN 978-93-5111-202-0

Published in 2014 in India by

RANDOM PUBLICATIONS

4376-A/4B, Gali Murari Lal, Ansari Road
New Delhi-110 002
Phone : +91-11-43580356, +91-11-23289044
e-mail: randomexports@gmail.com, sales@randompublications.com,
info@randompublications.com

Reprint 2021

Type Setting by : Keystoneprintads, Delhi-110051
Digitally Printed at: Replika Press Pvt. Ltd.

Preface

Soil quality is of fundamental importance for agricultural production, and soil fertility management is increasingly becoming a central issue in the decisions on food security, poverty reduction and environment management. For the purpose of safe ecological stewardship and achieving global food security, emphasising soil fertility management is becoming more and more important. Even so, the crucial role of soil fertility management for sustainable resource management and food security has been recognised only quite recently.

The goal of soil fertility management is to create soil chemical conditions that encourage plant growth and supply required nutrients in the amounts and at the times they are most needed. Liming materials and plant nutrients may be added to the soil in many forms and can be done so in a way that maximizes the economic benefits of nutrients while minimizing any environmental impact. The ways in which crops respond to these applications often are different because some soils have inherent physical limitations to plant growth.

Soil fertility is a complex process that involves the constant cycling of nutrients between organic and inorganic forms. As plant material and animal wastes decompose they release nutrients to the soil solution. Those nutrients may then undergo further transformations which may be aided or enabled by soil micro-organisms.

Natural processes such as lightning strikes may fix atmospheric nitrogen by converting it to (NO2). Denitrification may occur under anaerobic conditions (flooding) in the presence of denitrifying bacteria. The cations, primarily phosphate and potash, as well as many micronutrients are held in relatively strong bonds with the negatively charged portions of the soil in a process known as Cation Exchange Capacity.

Integrated Nutrient Management is a way of combating nutrient depletion. It is an approach to soil fertility management that combines organic and mineral methods of soil fertilisation with physical and biological measures for soil and water conservation. An agricultural ecosystem differs from a natural ecosystem in that plant nutrients are constantly being removed and exported. It is a holistic view of plant nutrient management, since it takes into consideration the totality of the farm resources that can be used as plant nutrients.

Soil Fertility Management for Sustainable Agriculture is an excellent reference for environmental and agricultural professionals as well as a book for undergraduate and graduate students preparing for a career in agriculture or soil fertility management.

– Author

Contents

Preface .... *v*

1. **Basic Characteristics of Soils** **1**

Origins of Soils from Rocks 1
Origin of Soils 3
Soil History: Deposition and Erosion 9
Soil Development 14
Soil Structure 18

2. **Organic Matter and Soil Structure** **27**

Organic Matter in Virgin and Cultivated Soils 29
Soil Microbial Activity 35
Limitations of Cover Crops 39
Organic Nitrogenous Fertilizers 42
Significance of Soil Structure 48
Soil Organic Matter on Soil Properties 51
Increased Soil Moisture 52
Nutrients in the Soil Organic Matter 55

3. **Mineral Function of Soils** **61**

Soil Minerals and Change in Mineral Composition 71
Minor Nutrients Levels in Soils 72
Nutrient Status of Soils Analysis 73
Essential Mineral Nutrients of Plant Growth on Soils 77

4. **Soil Biology and Nutrition** **81**

Soil Nutrients 84
Biological and Chemical Aspects of Soil Productivity 92
Importance of Soil Biology 96
Soil Components and Biological Processes 97

Soil Biological Properties 105
Soil Fungi 111
Soil Bacteria 114

5. Microorganisms of the Soil 117

Role and Application 120
Kinds of Soil Organisms 125
Functions of Microorganisms, Putrefaction, Fermentation, and Synthesis 127
Controlling the Soil Microflora 131
Soil Contamination on Soil Organisms (Diagnosis) 136

6. Basic Plant Nutrition 139

Essential Plant Nutrients 139
Soil Properties that Affect Plant Nutrition 143
Soil Microorganisms on Plant Health and Nutrition 144
Transformations that Supply Plant Available Nitrogen 148
Elements of Complete Plant Nutrition 151
A (Plant) Balanced Diet 154
Water and Mineral Uptake 157
Xylem and Transport 158
Essential Nutrients in Plants 163
Plant Effects on Soil Physical Condition 174

7. Soil Management 180

Soil Analysis 185
Soil Management Practices 188

8. Soil Fertility and Soil Mineralogy 194

Cation Exchange Capacity (CEC) and Anion Exchange Capacity (AEC) 194
Fertilizers and Nutrient Mining in Soils 198
Maintenance of Soil Fertility 200

9. Soil and Water Conservation Measures 205

Soil Water Relations under Field Conditions 210
Laboratory Measurements of Soil Water Relations 212
Soils and Soil Water Relations 213
The Effect of Water on Soil 216
Intensive Irrigation and Water Logging 218
Development of Tension in the Water Columns 225
Rates of Movement of Water through Plants 226

10. Crop Nutrition and Soils 227

Nutritive Value of Crop Residues 227
Rotations and Crop Sequence 231

Rotational Impacts 236
Factors Determining the Crop Coefficient 237
Soil Structure and Crop Growth 245
Proper Soil Sampling Equipment 248
Soil Moisture and Moisture Use 253

Index **257**

1

Basic Characteristics of Soils

Soils consist of grains (mineral grains, rock fragments, etc.) with water and air in the voids between grains. The water and air contents are readily changed by changes in conditions and location: soils can be perfectly dry (have no water content) or be fully saturated (have no air content) or be partly saturated (with both air and water present). Although the size and shape of the solid (granular) content rarely changes at a given point, they can vary considerably from point to point.

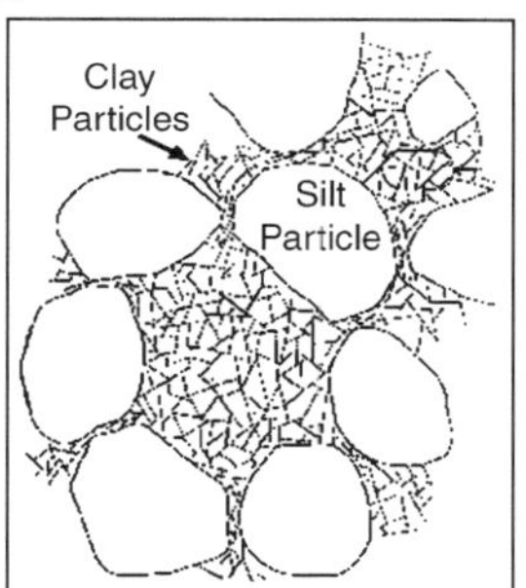

First of all, consider soil as a engineering material-it is not a coherent solid material like steel and concrete, but is a particulate material. It is important to understand the significance of particle size, shape and composition, and of a soil's internal structure or fabric.

ORIGINS OF SOILS FROM ROCKS

All soils originate, directly or indirectly, from solid rocks in the Earth's crust:

- *Igneous Rocks*: Crystalline bodies of cooled magma *e.g.* granite, basalt, dolerite, gabbro, syenite, porphyry
- *Sedimentary Rocks*: Layers of consolidated andcemented sediments, mostly formed in bodies of water (seas, lakes, etc.) *e.g.* limestone, sandstones, mudstone, shale, conglomerte
- *Metamorphic Rocks*: Formed by the alteration of existing rocks due to heat from igneous intrusions (*e.g.* marble, quartzite, hornfels) or pressure due to crustal movement (*e.g.* slate, schist, gneiss).

WEATHERING OF ROCKS

Physical Weathering

Physical or mechanical processes taking place on the Earth's surface, including the actions of water, frost, temperature changes, wind and ice; cause disintegration and wearing. The products are mainly coarse soils (silts, sands and gravels). Physical weathering produces Very Coarse soils and Gravels consisting of broken rock particles, but Sands and Silts will be mainly consists of mineral grains.

Chemical Weathering

Chemical weathering occurs in wet and warm conditions and consists of degradation by decomposition and/or alteration. The results of chemical weathering are generally fine soils with separate mineral grains, such as Clays and Clay-Silts. The type of clay mineral depends on the parent rock and on local drainage. Some minerals, such as quartz, are resistant to the chemical weathering and remain unchanged.

Quartz

A resistant and enduring mineral found in many rocks (*e.g.* granite, sandstone). It is the principal constituent of sands and silts, and the most abundant soil mineral. It occurs as equidimensional hard grains.

Haematite

A red iron (ferric) oxide: resistant to change, results from extreme weathering. It is responsible for the widespread red or pink colouration in rocks and soils. It can form a cement in rocks, or a duricrust in soils in arid climates.

Micas

Flaky minerals present in many igneous rocks. Some are resistant, *e.g.* muscovite; some are broken down, *e.g.* biotite.

Clay Minerals

These result mainly from the breakdown of feldspar minerals. They are very flaky and therefore have very large surface areas. They are major constituents of clay soils, although clay *soil* also contains silt sized particles.

CLAY MINERALS

Clay minerals are produced mainly from the chemical weathering and decomposition of feldspars, such as orthoclase and plagioclase, and some micas. They are small in size and very flaky in shape. The key to some of the properties of clay soils, *e.g.* plasticity, compressibility, swelling/shrinkage potential, lies in the structure of clay minerals.

There are three main groups of clay minerals:

1. *Kaolinites:* (Include kaolinite, dickite and nacrite) formed by the decomposition of orthoclase feldspar (*e.g.* in granite); kaolin is the principal constituent in china clay and ballclay.
2. *Illites:* (Include illite and glauconite) are the commonest clay minerals; formed by the decomposition of some micas and feldspars; predominant in marine clays and shales.
3. *Montmorillonites:* (Also called smectites or fullers' earth minerals) (include calcium and sodium momt-morillonites, bentonite and vermiculite) formed by the alteration of basic igneous rocks containing silicates rich in Ca and Mg; weak linkage by cations (*e.g.* Na^+, Ca^{++}) results in high swelling/shrinking potential.

ORIGIN OF SOILS

The constituents of the soil, like those of the plant, may be divided into the great classes of organic and inorganic. The origin of the former has been already discussed: they are derived from the decay of plants which have already grown upon the soil, and which, in various stages of decomposition, form the numerous class of substances grouped together under the name of humus. The organic substances may therefore be considered as in a manner secondary constituents of the soil, which have been accumulated in it as the consequence of the growth and decay of successive generations of plants, while the primeval soil consisted of inorganic substances only. The inorganic constituents of the soil are obtained as the result of a succession of chemical changes going on in the rocks which protrude through the surface of the earth.

We have only to examine one of these rocks to observe that it is constantly undergoing a series of important changes. Under the influence of air and moisture, aided by the powerful agency of frost, it is seen to become soft, and gradually to disintegrate, until it is finally converted into an uniform powder, in which the structure of the original rock is with difficulty, if at all distinguishable. The rapidity with which these changes take place is very variable; in the harder rocks, such as granite and mica slate it is so slow as to be scarcely perceptible, while in others, such as the shales of the coal formation, a very few years' exposure is sufficient for the purpose. These actions, operating through a long series of years, are the source of the inorganic constituents of all soils. Geology points to a period at which the earth's surface must have been altogether devoid of soil, and have consisted entirely of hard crystalline rocks, such as granite and trap, by the disintegration of which, slowly proceeding from the creation down to the present time, all the soils which now cover the surface have been formed. But they have been produced by a succession of very complicated processes; for these disintegrated rocks being washed away in the form of fine mud, or at least of minute particles, and being deposited at the bottom of the primeval seas, have there hardened into what are called sedimentary rocks, which being raised above the surface by volcanic action or other great geological forces, have been again disintegrated to yield different soils.

Thus, then, all soils are directly or indirectly derived from the crystalline rocks, those overlying them being formed immediately by their decomposition, while those found above the sedimentary rocks may be traced back through them to the crystalline rocks from which they were originally formed. Such being the case, the composition of different soils must manifestly depend on that of the crystalline rocks from which they have been derived. Their number is by no means large, and they all consist of mixtures in variable proportions of quartz, felspar, mica, hornblende, augite, and zeolites.

With the exception of quartz and augite, these names are, however, representatives of different classes of minerals. There are, for instance, several different minerals commonly classified under the name of felspar, which have been distinguished by mineralogists by the names of orthoclase, albite, oligoclase, and labradorite; and there are at least two sorts of mica, two of hornblende, and many varieties of zeolites. Quartz consists of pure silica, and when in large masses is one of the most indestructible rocks. It occurs, however, intermixed with other minerals in small crystals, or irregular fragments, and forms the entire mass of pure sand. The four kinds of felspar which have been already named are compounds of silica with alumina, and another base which is either potash, soda, or lime.

It is obvious that soils produced by the disintegration of these minerals must differ materially in quality. Those yielded by orthoclase must generally abound in potash, while albite and labradorite, containing little or none of that element must produce soils in which it is deficient. The quality of the soil they yield is not however entirely dependent on the nature of the particular felspar which yields it, but is also intimately connected with the extent to which the decomposition has advanced. It is observed that different felspars undergo decomposition with different degrees of rapidity but after a certain time they all begin to lose their peculiar lustre, acquire a dull and earthy appearance, and at length fall into a more or less white and soft powder.

During this change water is absorbed, and, by the decomposing action of the air, the alkaline silicate is gradually rendered soluble, and at length entirely washed away, leaving a substance which, when mixed with water, becomes plastic, and has all the characters of common clay. In this instance the decomposition of the felspar had reached its limit, a mere trace of potash being left, but if taken at different stages of the process, variable proportions of that alkali are met with. This decomposition of felspar is the source of the great deposits of clay which are so abundantly distributed over the globe, and it takes place with nearly equal rapidity with potash and soda felspar.

It is rarely complete, and the soils produced from it frequently contain a considerable proportion of the undecomposed mineral, which continues for a long period to yield a supply of alkalies to the plants which grow on them. Mica is a very widely distributed mineral, and two varieties of it are distinguished by mineralogists, one of which is characterised by the large quantity of magnesia it contains. Mica undergoes decomposition with extreme slowness, as is at once illustrated by the fact that its shining scales may frequently be met with entirely unchanged in the soil. Its persistence is dependent on the small quantity of alkaline constituents which it contains; and for this reason it is observed that the magnesian micas undergo decomposition less rapidly than those containing the larger quantity of potash.

Eventually, however, both varieties become converted into clay, their magnesia and potash passing gradually into soluble forms. Hornblende and augite are two widely distributed minerals, which are so similar in composition and properties that they may be considered together. In these minerals alkalies are entirely absent, and their decomposition is due to the presence of protoxide of iron, which readily absorbs oxygen from the air, when the magnesia is separated and a ferruginous clay left.

The minerals just referred to, constitute the great bulk of the mountain masses, but they are associated with many others which take part in the formation of the soil. Of these the most important are the zeolites which do not occur in large masses but are disseminated through the other rocks in small quantity. They are chiefly characterized by containing their silica in a soluble state, and hence may yield that substance to the plants in a condition particularly favourable for absorption. It is obvious from what has been stated that all these minerals are capable, by their decomposition, of yielding soft porous masses having the physical properties of soils, but most of them would be devoid of many essential ingredients, while not one of them would yield either phosphoric acid, sulphuric acid, or chlorine.

It has, however, been recently ascertained that certain of these minerals, or at least the rocks formed from them, contain minute, but distinctly appreciable traces of phosphoric acid, although in too small quantity to be detected by ordinary analysis; and small quantities of chlorine and sulphuric acid may also in most instances be found. Still it will be observed that most of these minerals would yield a soil containing only two or three of those substances, which, as we have already learned, are essential to the plant. Thus, potash felspar, while it would give abundance of potash, would be but an inefficient source of lime and magnesia; and labradorite, which contains abundance of lime, is altogether deficient in magnesia and potash. Nature has, however, provided against this difficulty, for she has so arranged it that these minerals rarely occur alone, the rocks which form our great mountain masses being composed of intimate mixtures of two or more of them, and that in such a manner that the deficiencies of the one compensate those of the other.

We shall shortly mention the composition of these rocks. Granite is a mixture of quartz, felspar, and mica in variable proportions, and the quality of the soil it yields depends on whether the variety of felspar present be orthoclase or albite. When the former is the constituent, granite yields soils of tolerable fertility, provided their climatic conditions be favourable; but it frequently occurs in high and exposed situations which are unfavourable to the growth of plants. Gneiss is a similar mixture, but characterised by the predominance of mica, and by its banded structure.

Owing to the small quantity of felspar which it contains, and the abundance of the difficulty decomposable mica, the soils formed by its disintegration are generally inferior. Mica slate is also a mixture of quartz, felspar, and mica, but consisting almost entirely of the latter ingredient, and consequently presenting an extreme infertility. The position of the granite, gneiss, and mica slate soils in this country is such that very few of them are of much value; but in warm climates they not unfrequently produce abundant crops of grain. Syenite is a rock similar in composition to granite, but having the mica replaced by hornblende, which by its decomposition yields supplies of lime and magnesia more readily than they can be obtained from the less easily disintegrated mica.

For this reason soils produced from the syenitic rocks are frequently possessed of considerable fertility. The series of rocks of which greenstone and trap are types, and which are very widely distributed, differ greatly in composition from those already mentioned. They are divisible into two great classes, which have received the names of diorite and dolerite, the former a mixture of albite and hornblende, the latter of augite and labradorite, sometimes with considerable quantities of a sort of oligoclase containing both soda and lime, and of different kinds of zeolitic minerals.

Generally speaking, the soils produced from diorite are superior to those from dolerite. The albite which the former contains undergoes a rapid decomposition, and yields abundance of soda along with some potash, which is seldom altogether wanting, while the hornblende supplies both lime and magnesia. Dolerite, when composed entirely of augite and labradorite, produces rather inferior soils; but when it contains oligoclase and zeolites, and comes under the head of basalt, its disintegration is the source of soils remarkable for their fertility; for these latter substances undergoing rapid decomposition furnish the plants with abundant supplies of alkalies and lime, while the more slowly decomposing hornblende affords the necessary quantity of magnesia. In addition to these, the basaltic rocks are found to contain appreciable quantities of phosphoric acid, so that they are in a condition to yield to the plant almost all its necessary constituents. The different rocks now mentioned, with a few others of less general distribution, constitute the whole of our great mountain masses; and while their general composition is such as has been stated, they frequently contain disseminated through them quantities of other minerals which, though in trifling quantity, nevertheless add their quota of valuable constituents to the soils. Moreover, the exact composition of the minerals of which the great masses of rocks are composed is liable to some variety. Those which we have taken as illustrations have been selected as typical of the minerals; but it is not uncommon to find albite containing 2 or 3 per cent of potash, labradorite with a considerable proportion of soda, and zeolitic minerals containing several per cent of potash, the presence of which must of course considerably modify the properties of the soils produced from them. They are also greatly affected by the mechanical influences to which the rocks are exposed; and being situated for the most part in elevated positions, they are no sooner disintegrated than they are washed down by the rains. A granite, for instance, as the result of disintegration, has its felspar reduced to an impalpable powder, while its quartz and mica remain, the former entirely, the latter in great part, in the crystalline grains which existed originally in the granite. If such a disintegrated granite remains on the spot, it is easy to see what its composition must be; but if exposed to the action of running water, by which it is washed away from its original site, a process of separation takes place, the heavy grains of quartz are first deposited, then the lighter mica, and lastly the felspar.

Thus there may be produced from the same granite, soils of very different nature and composition, from a pure and barren sand to a rich clay formed entirely of felspathic debris. The sedimentary or stratified rocks are formed of particles carried down by water and deposited at the bottom of the primeval seas from which they have been upheaved in the course of geological changes. The process of their formation may be watched at the present day at the mouths of all great rivers, where a delta composed of the suspended matters carried down by the waters is slowly formed. The nature of these rocks must therefore depend entirely on that of the country through which the river flows.

If its course runs through a country in which lime is abundant, calcareous rocks will be deposited, and if it passes through districts of different geological characters the deposit must necessarily consist of a mixture of the disintegrated particles of the different rocks the river has encountered. For this reason it is impossible to enter upon a detailed account of their composition.

It is to be observed, however, that the particles of which they are composed, though originally derived from the crystalline rocks, have generally undergone a complex series of changes, geology teaching that, after deposition, they may in their turn undergo disintegration and be carried away by water, to be again deposited. Their composition must therefore vary not merely according to the nature of the rock from which they have been formed, but also according to the extent to which the decomposition has gone, and the successive changes to which they have been exposed. They may be reduced to the three great classes of clays, including the different kinds of clay slates, shales, etc., sandstone and limestone.

It must be added also, that many of them contain carbonaceous matters produced by the decomposition of early races of plants and animals, and that mixtures of two or more of the different classes are frequent. The purest clays are produced by the decomposition of felspar, but almost all the crystalline rocks may produce them by the removal of their alkalies, iron, lime, etc. Where circumstances have been favourable, the whole of these substances are removed, and the clay which remains consists almost entirely of silica and alumina, and yields a soil which is almost barren, not merely on account of the deficiency of many of the necessary elements of plants, but because it is so stiff and impenetrable that the roots find their way into it with difficulty.

The sandstones are derived from the siliceous particles of granite and other rocks, and consist in many cases of nearly pure silica, in which case their disintegration produces a barren sand, but they more frequently contain an admixture of clay and micaceous scales, which sometimes form a by no means inconsiderable portion of them. Such sandstones yield soils of better quality, but they are always light and poor. Where they occur interstratified with clays, still better soils are produced, the mutual admixture of the disintegrated rocks affording a substance of intermediate properties, in which the heaviness of the clay is tempered by the lightness of the sandstone. Limestone is one of the most widely distributed of the stratified rocks, and in different localities occurs of very different composition.

Limestones are divided into two classes, common and magnesian; the former a nearly pure carbonate of lime, the latter a mixture of that substance with carbonate of magnesia. But while these are the principal constituents, it is not uncommon to find small quantities of phosphate and sulphate of lime, which, however trifling their proportions, are not unimportant in an agricultural point of view. These limestones are hard and possess to a greater or less extent a crystalline texture. They are replaced in later geological periods by others which are much softer, and often purer, of which the oolitic limestones, so called from their resemblance to the roe of a fish, and chalk are the most important. Other limestones are also known which contain an admixture of clay. The soils produced by the disintegration of limestone and chalk are generally light and porous, but when mixed with clay, possess a very high degree of fertility, and this is particularly the case with chalk, which yields some of the most valuable of all soils. But it is true only of the common limestones, for experience has shown that those which contain magnesia in large quantity are often prejudicial to vegetation, and sometimes yield barren or inferior soils. Such are the general characters of the three great classes of stratified rocks; any attempt to particularise the numerous varieties of each would lead us far beyond the limits of the present work. It is necessary, however, to remark, that in many instances one variety passes into the other, or, more correctly speaking, sedimentary rocks occur, which are mixtures of two or more of the three great classes.

In fact, the name given to each really expresses only the preponderating ingredient, and many sandstones contain much clay, shales and clay slates abound in lime, and limestones in sand or clay, so that it may sometimes be a matter of some difficulty to decide to which class they belong. Such mixtures usually produce better soils than either of their constituents separately, and accordingly, in those geological formations in which they occur, the soils are generally of excellent quality. The same effect is produced where numerous thin beds of members of the different classes are interstratified, the disintegrated portions being gradually intermixed, and valuable soils formed. The fertility of the soils formed from the stratified rocks is also increased by the presence of organic remains which afford a supply of phosphoric acid, and which are sometimes so abundant as to form a by no means unimportant part of their mass. They do not occur in the oldest sedimentary rocks, but as we ascend to the more recent geological epochs, they increase in abundance, until, in the greensands and other recent formations, whole beds of coprolites and other organic remains are met with. Great differences are observed in the quality of the soils yielded by different rocks. In general, those formed by the disintegration of clay slates are cold, heavy, and very difficult and expensive to work; those of sandstone light and poor, and of limestone often poor and thin. These statements must, however, be considered as very general; for individual cases occur in which some of these substances may produce good soils, remarkable exceptions being offered by the lower chalk and some of the shales of the coal formation.

Little is at present known regarding the peculiar nature of many of these rocks, or their composition; and the cause of the differences in the fertility of the soil produced from them is a subject worthy of minute investigation.

SOIL HISTORY: DEPOSITION AND EROSION

ORIGINAL DEPOSITION

Most soils are formed in layers or lenses by deposition from moving water, ice or wind. One-dimensional compression occurs as overlying layers are added. Vertical and horizontal stresses increase with deposition.

Erosion

Erosion causes unloading; stresses decrease; some vertical expansion occurs. Plastic strain has occurred; the soil remains compressed, *i.e.* overconsolidated.

Subsequent Changes

Subsequent changes may occur in the depositional environment: further loading/unloading due to glaciation, land movement, engineering; and ageing processes.

AGEING

The term *ageing* includes processes that occur with time, except loading and unloading. Ageing processes are independent of changes in loading.

Vibration and Compaction

Coarse soils can be made more dense by vibration or compaction at essentially constant effective stress

Creep

Fine soils creep and continue to compress and distort at constant effective stress after primary consolidation is complete.

Cementing and Bonding

Intergranular cementing and bonding occurs due to deposition of minerals from groundwater, *e.g.* calcium carbonate; disturbance due to excavation fractures the bonding and reduces strength.

Weathering

Physical and chemical changes take place in soils near the ground surface due to the influence of changes in rainfall and temperature.

Changes in Salinity

Changes in the salinity of groundwater are due to changes in relative sea and land levels, thus soil originally deposited in sea water may later have fresh water in its pores, such soils may be prone to sudden collapse.

DENSITY INDEX (RELATIVE DENSITY)

The void ratio of coarse soils (sands and gravels) varies with the state of packing between the loosest practical state in which it can exist and the densest. Some engineering properties are affected by this,

Shear Strength

Seepage

The flow of water through soil.

Seepage Force (J)

The force transmitted to a body of soil due to the seepage of groundwater. $J = i\,\gamma_w\,V$

Seepage Pressure (j)

The seepage force per unit volume $j = i\,\gamma_w$

Seepage Velocity (v_s)

The average velocity at which groundwater flows through the pores; the ratio of the volume flow rate to the average area of voids in a cross-section.

$$v_s = q/\,A_v$$

Sensitivity (S_t)

$$S_t = \frac{\text{Undisturbed undrained strength}}{\text{Remoulded undrained strength}}$$

A measure of the change in strength of f clays upon disturbance: For ordinary clays S_t = 1 to 4, sensitive clays 4 to 8, 'quick' clays 16–100.

Settlement (s, ρ)

The downward movement of ground or ground surface; the downward movement of a foundation.

Shape Factors (s_c s_q s_g)

Factors used in a general equation giving ultimate bearing capacity which provide adjustment relating to the shape (*e.g.* strip, square, circle)

Shear Modulus (G')

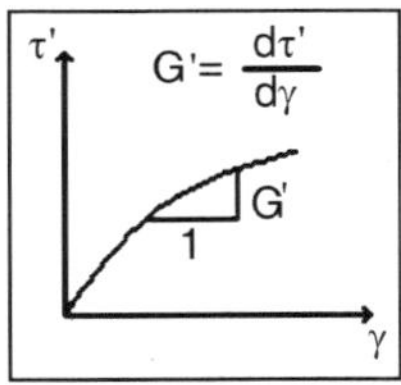

The ratio of the change in shear stress to the resulting change in shear strain.

Shear Strain (γ)

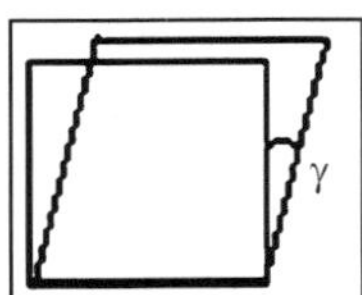

The angular distortion or change in shape of a body of material.

Shear Strength (τ_f)

The maximum shear stress which a material can sustain under a given set of conditions. In soil mechanics it is necessary to refer shear strength to the strain at which the strength is measured.

- *Critical shear strengt*: $\tau_c = c'_c + \sigma' \tan\phi'_c$
- *Peak shear strength*: $\tau_p = c'_p + \sigma' \tan\phi'_p$
- *Undrained shear strength*: $\tau_c = s_u$
- *Residual shear strength*: $\tau_r = c'_r + \sigma' \tan\phi'_r$

Shear Stress (τ)

The force per unit area acting tangentially to a given plane or surface. Units: kPa.

Short-term Conditions

Conditions in the ground when it is undrained; loading or unloading will cause changes in pore pressures, but not immediately in volume; these excess pore pressures dissipate with time due to consolidation.

Shrinkage Limit (w_s)

(Also SL) The water content below which further reduction in water content causes no further reduction in volume.

Skin Friction Stress (f_s)

The shear stress on the shaft of a pile of a pile or caisson or cone penetrometer. *Note:* Although known as 'skin friction', the shear stress may not vary with normal stress.

Soil Suction

Negative pore pressure created by capillary attraction in fine soils and in unsaturated soils.

Specific Gravity (G_s)

The ratio of the mass of a body or a substance to the mass of an equal volume of water; the ratio of the density of a body or a substance to that of water.

$$G_s = \frac{\text{Mass of body}}{\text{Mass of the same volume of water}}$$

Specific Surface (S_s)

The total surface area of all particles in a unit mass of soil. Units: m^2/g.

Specific Volume (v)

The total volume of a quantity of soil containing a unit volume of soil grains.

$$v = 1 + e$$

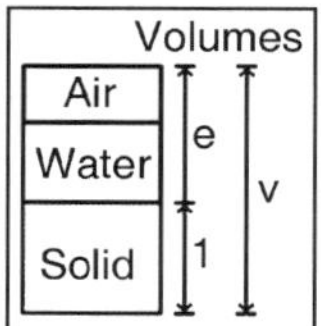

State Boundary Surface

The state boundary surface (SBS) is the boundary (usually in q′:p′:v space) to all possible stress-volume states of a soil. If a soil with a state on the state boundary surface is unloaded, the subsequent state will be inside the SBS, upon re-loading the subsequent state will move back onto (but not beyond) the SBS. In other words, stable states cannot exist outside the SBS.

State Parameter (S_s, S_v)

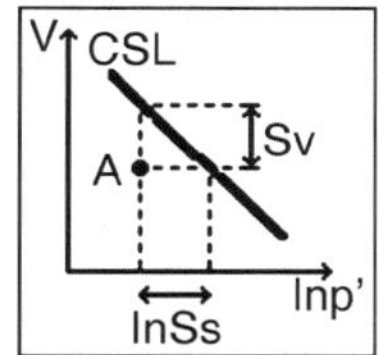

A measure of the distance between the current state (A) and the critical state; expressed as a ratio of stresses or as a difference of specific volumes,

$$S_s = p'_A / p'_c$$

$$S_v = v_A - v_c$$

Steady State Pore Pressure (u_0)

The pore pressure at equilibrium when all excess pore pressures have fully dissipated.

Stiffness

Susceptibility to distortion or volume change under load.

Strain

A measure of the change in size or shape of a body, relative to its original size or shape. (Direct strain is the ratio of change in length to original length; shear strain is the angle of distortion; volumetric strain is the ratio of change in volume to original volume.)

Stream Function (Ψ)

A function introduced in the solution of the Laplace equations defining two-dimensional seepage flow; the flow-quantity interval ($\Delta\theta$) between stream lines (flow lines) can be stated as $\Delta\Psi = \Delta\theta$.

Stress

The intensity of force per unit area; normal stress is applied perpendicularly to a surface or plane, shear stress is applied tangentially to a surface or plane.

Stress History

The past history of loading and unloading associated with a soil.

Summation (Σ)

The symbol Σ when placed in front of a quantity indicates that the quantity is a sum or total.

Swelling/Recompression Index (C_s, k)

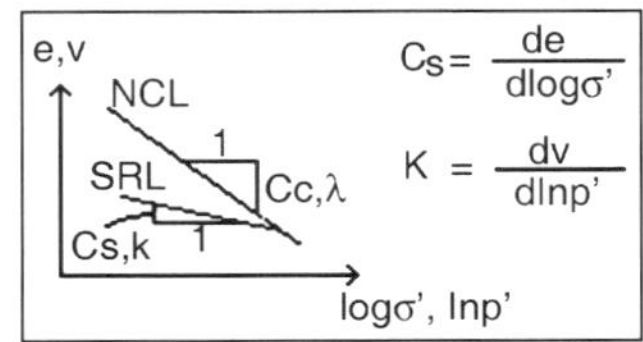

The slope of the swelling (unloading) and recompression (reloading)line. $C_s = 2.303$ k where k = slope of v:ln σ′ swelling/recompression line.

SOIL DEVELOPMENT

Soil forms from a complex interaction between earth materials, climate, and organisms acting over time. The brightly coloured soils of the humid tropics reflect the intense chemical reactions occurring in warm climates. The fertile prairie soils of the American Midwest evolved from the nutrient-rich organic matter left by decaying grasses. Regardless of soil characteristics, the whole process starts with the breakdown of earth material.

WEATHERING

Weathering refers to processes that physically breakdown and chemically alter earth material. *Physical weathering,* also known as *mechanical weathering,* is the breakdown of large pieces of earth material into smaller ones. Think of physical weathering as the *disintegration* of rock without changing its chemical composition. There are many ways earth material can be physically weathered. When water freezes in rock crevices it expands creating stress in the crevice. As the stress increases, the crevice widens ultimately breaking the rock. Plant roots wedge rocks apart as they grow into rock crevices too. The shrinking and swelling by alternating heating and cooling weakens mineral bonds causing the rock to disintegrate. A very important result of physical weathering is its impact on the surface area of weathered material. When a block of earth material is broken into several smaller pieces, the amount of exposed surface increases. Examine the diagram below. A block with a width, depth, and height of 1 cm has a total surface area of 6 square centimeters. If we break the block in half in all directions it yields eight smaller pieces all with width, height, and depth of.5 cm. Breaking the block apart creates additional exposed surfaces such that the total surface area is now 12 square centimeters. Having more total exposed surface provides more area upon which chemical reactions can take place to further weather the material. The shape of the pieces also affects the the amount of exposed surface area. Plate-like pieces have more exposed surface area than do block-like pieces.

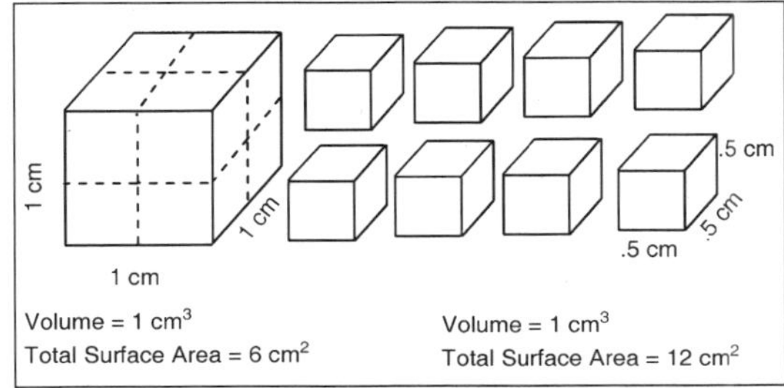

Fig. Effect of Physical Weathering on Surface Area

Chemical weathering breaks down earth material by chemical alteration. This usually means adding a substance like water or air to the material. For instance, when oxygen is added to iron bearing minerals, *oxidation* takes places and a loose mantle of iron oxide is created (rust). *Hydrolysis* is an exchange reaction involving minerals and water. Free hydrogen (H^+) and hydroxide $(OH)^-$ ions in water replace mineral ions and drive them into solution. As a result, the mineral's structure is changed into a new form. Hydrolysis is a common process whereby silicate minerals are weathered into a clay mineral. Think of chemical weathering as the decomposition of earth material.

The spatial variation of climate and organisms play a significant role in the weathering of earth materials. Dry locations tend to be dominated by physical weathering and moist places by chemical weathering. The type of earth material available also determines the amount of weathering that might take place. Limestone is easily broken down where abundant rainfall and high temperatures prevail. However, limestone will remain intact in dry locations. The end result of the weathering process is the creation of a *weathered mantle.* The weathered mantle is not yet a soil until it undergoes further change. This involves the addition, transformation, translocation and removal of materials from the weathered mantle to form distinctive soil layers.

Since early geologic time, the atmosphere has interacted with the Earth's exposed crust though a process known as *weathering*. Weathering takes place through a combination of both mechanical and chemical means. We have all experienced the results of weathering first hand. Any visit to an old cemetery find us peering at the blurred inscriptions on old marble tombstones. These inscriptions were once perfectly legible, but with the passage of time, the small fractures and cracks in the rock have made it vulnerable to attack by aqueous solutions. A dramatic example of weathering can be seen in these two images of the same 3000 year old Egyptian obelisk just before relocation to damp New York and 100 years after accelerated weathering in New York Weathering rates are obviously a strong function of climate!

Many of the original volcanic gases (*e.g.*, carbon dioxide, sulphur-bearing gases, etc.) were able to dissolve in water and produce acids. The acids, in turn, reacted with surface minerals.

Later, oxygen in the atmosphere reacted with the exposed reduced materials, making the red beds discussed in an earlier lecture. Since the advent of land plants, soil and surface minerals have been exposed to relatively high concentrations of carbon dioxide maintained in soil pores as a result of decomposition and the metabolic activities of roots. The reaction of carbon dioxide with water in the soil produces carbonic acid (H_2CO_3) which determines the rate of rock weathering in most ecosystems.

$$CO_2 \text{ (gas)} + H_2O \text{ (liquid)} - H_2CO_3 \text{ (solution)}$$

Acid rain, produced by human effluents of nitrogen and sulphur-bearing gases will increase the rate of rock weathering in downwind areas. To understand why weathering occurs and why the rate of rock weathering is so dependent on climate, we need to discuss the chemistry of the process in a little more detail.

IGNEOUS ROCK WEATHERING

There is a well known expression that captures much of the story of rock weathering and illustrates the important role it has played in Earth history:

"Igneous Rocks + Acid Volatiles = Sedimentary Rocks + Salty Oceans"

What we mean by this will become clearer if we look at the details of the weathering processes. Consider a boulder or rock containing *Feldspar* minerals. Feldspar is a general term for a group of aluminosilicate minerals containing sodium, calcium, or potassium and having a lattice framework structure that makes for rigidity. Feldspars turn out to be one of the most common minerals in the Earth's crust. Feldspars are weathered through the chemical process of *hydration:*

$$K\,Al\,Si_3O_8 + H_2O \rightarrow Al_2SiO_5(OH)_4$$

In this chemical formula feldspar reacts with water to produce a kaolinite (clay). Notice that the chemical equation does not exactly balance, that is, not all the elements on the left hand side appear on the right hand side. This is because soluble elements, such as potassium (K) are *leached* out during the chemical reaction and carried away as dissolved salts. The process of leaching can perhaps best be understood by analogy with the making of coffee. When hot water is passed over crushed coffee beans, the soluble components (making the coffee) are leached away, leaving the insoluble crushed coffee bean remnants behind.

In this way, rocks containing feldspars are weakened though the conversion of rigid feldspar to more plastic clays which do not have anything like the same structural rigidity. The process occurs at exposed surfaces of the minerals making up the rock. Through geologic time, large amounts of sedimentary r0ocks have been deposited as part of this process. In fact about 75% of all exposed rocks on the Earth's surface today are of sedimentary origin and have been brought to the surface by geologic uplift.

Of course, geological processes return some of the sedimentary rocks to the mantle of the Earth, where they are converted back to the primary materials under conditions of great temperature and pressure.

Rock weathering is also critical for the release of biochemical elements that have no gaseous form-examples are calcium, Ca, Potassium, K, Iron, Fe, and Phosphorus, P. The latter element plays a key role in cell metabolism. Thus, we can say that weathering provides key nutrients for life through the process of leaching.

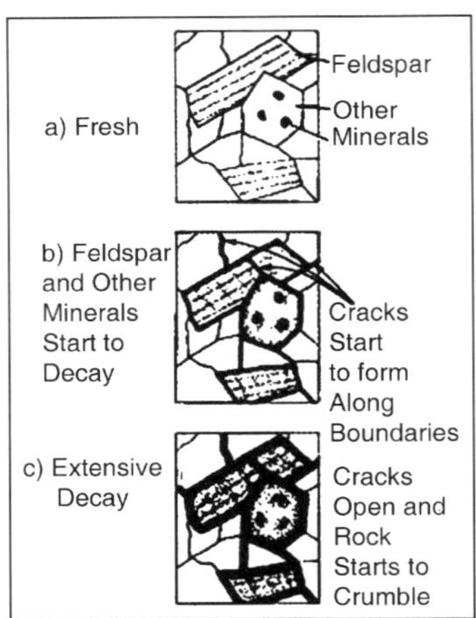

The soluble nutrients are transferred to soils layers below the immediate surface (*e.g.*, the "B" layer). This is a major reason that plants have evolved root systems-to search out these critically needed nutrients below ground.

Igneous rock weathering proceeds in stages. It is useful to picture the inside of a rock-made up of interlocking minerals of irregular shapes, each being rigid. Figure shows a microscopic view of such an interior rock composition, with some of the grains being feldspars.

As weathering proceeds, the boundaries of the vulnerable feldspar (and other) mineral grains start to decay. As the decay proceeds, water can reach more and more feldspar surfaces and the process accelerates. This process can also be accelerated by melt-freeze cycles that force the grains apart due to the difference between the volume occupied by water and ice. We can see that weathering is due to the combined effects of chemical and mechanical decay.

The chemical and physical processes of weathering transform the igneous rock into sand and clay particles and dissolved salts. Chemical weathering can add carbon dioxide, water, and oxygen. The link provided courtesy of the National Park Service shows an example of a weathering rock. It is interesting to note that, since most of the Earth's exposed rock is of sedimentary origin and since sedimentary rocks are a by-product of weathering, most of the rocks that are weathering away in today's world are second or perhaps even third generation rocks.

That is, they originated as igneous material, became sedimentary through weathering and transport to the bottom of shallow waters, were then subject to geologic uplift to become again exposed and, finally, began to undergo weathering yet again. The natural world is full of such endless cycles.

SOIL STRUCTURE

Soil stricture refers to the arrangement of soil particles, both primary and secondary. Soil structure is one of the most important properties of soil mass, since, it influences aeration, permeability, water capacity etc. In the field the structure is described in terms of:

- Type-referring to shape and arrangement
- Class-referring to size
- Grade-referrint to the degree of aggregation.

Types of Soil

There are 4 types of primary structures:

1. *Platy*: With particles arranged around a plane, generally horizontal.
2. *Prism-like*: With particles arranged around a vertical axis and bounded by relatively flat vertical surfaces which are the casts or the molds formed by the faces of the surrounding peds. This type includes prismatic and columnar types.
3. *Block-like*: With particles arranged around a point and bounded by flat or rounded surfaces which are the casts or the molds formed by the faces of the surrounding peds. This type includes angular and subangular types.
4. *Spheroidal*: Particles arranged around a point and bounded by curved or very irregular surfaces that are not accommodated in the adjoining aggregates. This type includes granular and crumb types.

Classes

Five size classes are recognized in each of the primary types. They are fine, very fine, medium, coarse and very coarse. The actual size for classes in each type varies.

Grade

The grade of structure representing the degree of aggregation is determined in the field mainly by noting the durability of the aggregates and the proportions between aggregated and unaggregated material that results when the aggregates are disturbed or gently crushed. Grades are termed structureless (massive if coherent, and single-grained if non-coherent), weak, moderate, strong and very strong, depending on the stability of aggregates when disturbed.

Factors Affecting Structure.

Structure is primarily influenced by texture. The beneficial effects of organic matter on the structure and working of the soils are well known.

Soil structure is also influenced by:

- *Soil Management*: The cutting action of ploughs or other tillage implements breaks up the soil mass and may have favourable or adverse effects on the structure, depending on whether the soil is worked under optimum moisture conditions or not. A good soil management, with proper system of crop rotation has the effect of maintaining the soil in a good state of aggregation.

- *Absorbed Cations*: Sodium and potassium ions on the clay complex have a tendency to disperse the soil; calcium has favourable effects on the aggregation. Similarly the presence of soluble salts favours floculation.
- *Micro-Organisms*: The filamentous growth of soil fungi and the microbial decomposition products of organic matter have a binding effect on soil particles thus favouring aggregation. Similarly the excrements of earthworms and the burrowing activities of insects and other creatures cause appreciable changes in the soil structure.
- *Variations of Soil Moisture*: Variations of soil moisture due to drying and wetting influence the structure. The drying of soil forms cracks and big clods. Poorly drained soils with excess of moisture usually have an unfavourable structure.

The structure of cultivated soils have a very important bearing on their drainage, ease of tillage, root penetration and resistance to erosion and ultimately their general productivity. From this point of view the crumb and granular structure (spheroidal) is considered very favourable to plant growth.

The structure and physical properties of soil can be improved by adopting a suitable system of soil management, including legumes in the rotation system, green manuring, and regularly supplying the soil with organic manure.

Mineral Composition of the Soil Separates

Owing to a varying rate if wearing down of different minerals, the more easily decomposing feldspars predomonate in the relatively fine separates, whereas the resistant minerals, *e.g.*, are more abundant in coarser separates. The separates therefore show differences in chemical composition also. As indicated earlier, the clay which has a high specific surface is more reactive than silt or sand.

The Clay Fraction

In the process of decomposition relatively stable new elements are formed from the products of weathering which constitute largely the clay fraction. The clay particles formed in this manner are crystalline and as shown by X-ray diffraction and ptrographic techniques are composed of sheets of hydrated alumina and silica linked by oxygen atoms. Because of their being very small, they are highly reactive and form the seat of ion exchange in the soil and this controls and regulates adsorption, retention, and release of many plant nutrients, such as potassium, calcium, magnesium and phosphorous. The precise nature of clay properties depends on the type of minerals that predominantly compose the clay.

These fall into three main groups:

1. *Kaolinite*: A unit kaolinite crystal lattice consists of one sheet of silica and one sheet of alumina and is hence, called 1:1-layer silicate. The two sheets are held together by mutually shared oxygen atoms. These units in turn are tenaciously held together by oxygen-hydroxyl linkages and consequently no expansion occurs when the mineral is wetted. It has a low specific surface and a low caton exchange capacity. Similarly plasticity, cohesion, shrinkage and the swelling properties of kaolinite are low.
2. *Montmorillonite*: A unit montmorillonite crystal lattice consists of 2 sheests of silica and 1 sheet of alumina (2:1-layer silicate) held together by mutually shared oxygen atoms. Such units however are held together by weak oxygen-oxygen linkages. There is some isomorphic substitution of iron or magnesium in the alumina sheet; mont morillonite crystals can expand and owing to this expansion, cations and water molecules are able to move in between the crystal units. Montmorillonite has a high specific surface and caton-exchange capacity. Similarly, plasticity, cohesion, shrinkage and sweeling properties are high.
3. *Illite*: Illite has the same general structure organisation as montmorillonite except in respect of the linkages between the crystal units. About 15% of silica in the silica sheet is replaced by aluminium and potassium atoms and they supply the additional connecting linkages between the crystal units, due to which illite shows a lower expansion capacity. It has properties intermediate between kaolinite and montmorillonite.

Soil Colloids

The most active portions of the soil are those which are in the colloidal state. The colloidal state implies a two\phase system in which the material called the dispersed phase (fine clay and humus) is dispersed in the dispersed medium (water). In soils, the mineral and organic colloids exist in intimate and heterogenous admixture.

The mineral colloid is present almost exclusively as the clay of various kinds, whereas the organic colloid is present as humus. It is generally supposed that clay particles less than one micron in diameter possess colloidal properties and these properties increase with a decrease in the size of the particles.

The most distinctive colloidal properties are:

- The large specific surface or interface, and
- The capacity to hold solids, gases, salts and ions.

Depending on their nature, the soil colloids determine many of the physical and chemical properties. The colloidal material may occur as a thin gelatinous film around coarser particles, or it may occupy a considerable part of the space between the larger particles, thus serving as a binding material.

Soil colloids especially the mineral colloids, exhibit cohesive and adhesive properties. Soils with a high amount of colloidal clay compete with plant roots for water and mineral nutrients, especially at lower levels of their availability. Generally the soil colloids have a high exchange capacity, which increases with silica sesquioxide ratio.

CHEMICAL PROPERTIES

Mineral Matter and its Composition

The principal minerals occurring in the earth's crust are:

Mineral	Approximate percentage
Feldspars	48
Quartz	36
Micas	10
Limestone and dolomite	2
Horneblende and augite	1
Clays	1
Other minerals	1

- *Feldspars*: Feldspars are anhydrous alumino silicates of potassium, sodium and calcium.
- *Quartz*: Quartz or silica is silicon dioxide found in crystalline rocks. It also occurs as free silica or sand.
- *Mica*: Mica is layer of alumino silicate of potassium; iron, magnesium and sodium may be present in varying proportions.
- *Limestone*: It is largely calcium carbonate; with magnesium carbonate it occurs as dolomite.
- *Hornblende and Augite*: They are the ferro magnesium minerals consisting of silicates of calcium, magnesium, iron and sodium in varying proportions.
- *Olivine and Serpentine*: Olivine is an olive-green ferro-magnesium silicate. Serpentine is a hydrated silicate of magnesium.
- *Clays*: Clays are secondary minerals. They are hydrated alumino silicates.
- *Other Minerals*: Other minerals occurring in the soil include tourmaline- a boro alumino silicate with alkali metals and iron or magnesium; rutile-titanium oxide, zircon-ziconium silicate,

glauconite-hydrated silicate of iron and potassium, apatite-calcium phosphate, sulphur-bearing minerals, etc. Allophane and the hydrous oxides of aluminium, iron and titanium also commonly occur in soils and more so in highly weathered and leached soils.

Inorganic Components

From what has been stated in the preceding section it is evident that compounds of silicon, aluminium, calcium, magnesium, iron, potassium and sodium form the principal chemical constituents of the mineral matter in the soils. Besides, the soil contains small amounts of a large number of other mineral elements, *e.g.*, phosphorus, boron, manganese, copper, sulphur, zinc and cobalt. Thus the soil supplies all the following essential mineral elements required by the plants.

- *Macros*: Phosphorus, potassium, calcium, magnesium and sulphur.
- *Micros*: Iron, manganese, zinc, copper, molybdenum, boron and chlorine.

Plants obtain carbon, hydrogen and oxygen from carbon dioxide and water. Soil is also the source of nitrogen for plants, but the ultimate source of nitrogen is traceable to atmosphere from where nitrogen fixation takes place physico-chemically and biologically.

The total amount of elements contained in the soils depends partly on the nature of the parent material from which they are formed and partly on their age and extent to which soluble products have been leached down. The chemical composition of different horizons of a soil also show a good deal of variation. Generally, 'A' horizon is richer in soluble components than the 'B' horizon. Usually some of the elements that are commonly leached out are also ones that are required by the plants tables give the percentage inorganic composition of eight representative Indian soils and os the successive horizons of the soils.

Ion Exchange

It is a reversible process by which cations and anions are exchanged between solid and liquid phases which are in close contact with each other. The exchange of cations and anions is termed cation exchange and anion exchange respectively. Ion exchange is the most important of all the processes occurring in the soil. Soil colloids are the seats of ion exchange. Chemical and physical processes connected with ion exchange include the weathering of minerals, nutrient absorption by plants, the swelling and shrinkage of clay, leaching of soluble salts, etc. The capacity of soils to adsorb and exchange cations and anions varies greatly with the nature and amount of clay, and the organic matter. The cation exchange capacity (CEC) is defined as the amount of a cation species bound at pH 7 (neytral pH) and is expressed as milli-equivalents per 100 grammes. The CEC values of reference clay minerals are:

Kaolinite	3-10 m.e./100 g
Illite	10-30 m.e./100 g
Montmorillonite	80-150 m.e./100 g

Organic colloids may have CEC of 200 m.e./100 g or more.

Base Saturation

The portion of CEC accounted for by the basic ions (calcium, magnesium, potassium and sodium) and expressed as percentage of the CEC is the base saturation percentage. A soil saturated with calcium and magnesium is considered normal and fertile. If a soil has more than 15% exchangeable sodium, it is an alkali soil. If the soils are base-unsaturated, *i.e.,* the proportion of exchangeable hydrogen is more, the soil tends to be acidic.

Acidic, saline and alkali soils are separately dealt with. Similarly other items, *viz.* pH and conductivity are discussed in the section 'Soil-Testing'.

ORGANIC MATTER

Organic matter though forming a small part of mineral soils plays a vital role in the productivity and conditioning of soils. It serves as source of food for soil bacteria and fungi which are responsible for converting complex organic materials into simple substances readily used by the plants. The intermediate products of decomposition of fresh organic matter help to increase the physical condition of the soil.

The addition of organic matter also improves the working quality or friability of the soil. In association with clay and calcium, it helps to form the aggregates of soil particles to produce the 'crumb structure'. Applied in adequate quantity organic matter serves as a mulch.

Humus

The organic matter in the soil consists largely of plant remains, the residues of soil-microorganisms feeding on them and several products of their decomposition. In consonance with its origin, organic matter contains most of the mineral elements found in the plants. The plant remains may occur in recognizable form but more commonly they are found as a fairly stable, dark, amorphous complex colloidal substance called humus.

Chemically, humus represents a mixture of decomposed or altered products of carbohydrates, proteins, fats, resins, wax and other similar substances.

These complex compounds are gradually decomposed by soil organisms into simple mineral salts, carbon dioxide, water, organic acids, ammonia, methane and free nitrogen, depending upon the initial composition of the organic matter. The average composition of humus is as follows;

	Percentage
Carbon	50
Oxygen	35
Nitrogen	5
Hydrogen	5
ash(containing phosphorus, potassium, sulphur and other elements)	5

Humus is soluble in water only to a very little extent but owing to high specific surface, it can absorb gases as well as water. Lime has a precipitating action on humus. Humus also exhibits exchange properties.

Content of Organic Matter

The content of organic matter varies with the kind of soil, vegetation, climate, cultivation, manurial and rotation practices, and biological activities.

The vegetation determines the quality and quantity of organic material added each year, whereas the climate determines the rate of decomposition. In the Indian soils the content of organic matter is generally low because of the high rate of decomposition under tropical and subtropical climate. Except in a few localised areas in the hilly regions, the organic matter in most if the cultivated soils rarely exceeds 1 per cent.

Soil group	**Percentage organic carbon**
Deep black soil	0.34-0.77
Red and laterite soil	0.68-6.53
Alluvial soil	0.28-1.10

The movement and distribution of humus in the soil depends on the pedogenic processes. Normally, the amount of organic matter and humus decreases from the surface downwards. Usually in the forest soils owing to continual leaf fall, the amount of organic matter is high. Under acidic conditions humus becomes dispersed and is carried down into the subsoil where it may form part of the hard accumulated layer of mineral and organic colloidal matter.

Carbon-Nitrogen Ratio

Nitrogen plays major role in the growth and reproductive process of plants. The organic combination of nitrogen with organic matter in the soil constitutes the principal storehouse from which nitrogen is slowly made available to the crops. The relation of organic matter to nitrogen is of particular interest. Investigations prove that ratios of nitrogen, carbon and organic matter in the soil are fairly constant. From the average composition of humus given already, the carbon: nitrogen ratio of organic matter is 10 to 1.

The general level of carbon and nitrogen in most of the Indian soils is low and the carbon: nitrogen ratios fluctuate widely from 5 to 25, with the average value around 14.

2

Organic Matter and Soil Structure

A major benefit obtained from green manures is the addition of organic matter to the soil. During the breakdown of organic matter by microorganisms, compounds are formed that are resistant to decomposition—such as gums, waxes, and resins. These compounds—and the mycelia, mucus, and slime produced by the microorganisms—help bind together soil particles as granules, or aggregates.

A well-aggregated soil tills easily, is well aerated, and has a high water infiltration rate. Increased levels of organic matter also influence soil humus. Humus—the substance that results as the end product of the decay of plant and animal materials in the soil—provides a wide range of benefits to crop production. Sod-forming grass or grass-legume mixtures are important in crop rotations because they help replenish organic matter lost during annual cultivation.

However, several years of sod production are sometimes required before measurable changes in humus levels occur. In comparison, annual green manures have a negligible effect on humus levels, because tillage and cultivation are conducted each year. They do replenish the supply of active, rapidly decomposing organic matter. The contribution of organic matter to the soil from a green manure crop is comparable to the addition of 9 to 13 tons per acre of farmyard manure or 1.8 to 2.2 tons dry matter per acre.

Nitrogen Production

Nitrogen production from legumes is a key benefit of growing cover crops and green manures. Nitrogen accumulations by leguminous cover crops range from 40 to 200 lbs. of nitrogen per acre. The amount of nitrogen available from legumes depends on the species of legume grown, the total biomass produced, and the percentage of nitrogen in the plant tissue. Cultural and environmental conditions that limit legume growth—such as a delayed planting date, poor stand establishment, and drought—will reduce the amount of nitrogen produced. Conditions that encourage good nitrogen production include getting a good stand, optimum soil nutrient levels and soil pH, good nodulation, and adequate soil moisture.

The portion of green-manure nitrogen available to a following crop is usually about 40 per cent to 60 per cent of the total amount contained in the legume. For example, a hairy vetch crop that accumulated 180 lbs. N per acre prior to plowing down will contribute approximately 90 lbs. N per acre to the succeeding grain or vegetable crop. Dr. Greg Hoyt, an agronomist at North Carolina State University, has estimated that 40 per cent of plant tissue nitrogen becomes available the first year following a cover crop that is chemically killed and used as a no-till mulch. He estimates that 60 per cent of the tissue N is released when the cover crop is incorporated as a green manure rather than left on the surface as a mulch. Lesser amounts are available for the second or third crop following a legume, but increased yields are apparent for two to three growing seasons.

To determine how much nitrogen is contained in a cover crop, an estimate is needed of the yield of above-ground herbage and its percentage of nitrogen. A procedure to make this determination is available in the *Northeast Cover Crop Handbook,* in *Farmer's Fertilizer Handbook,* and in *Managing Cover Crops Profitably*. The procedure involves taking a field sample, drying it, weighing it, and sending a sample off for forage analysis, which includes an estimate of protein content. Once the protein content is known, simply divide it by 6.25 to obtain the percentage of nitrogen contained in the cover crop tissue.

WHAT IS ORGANIC MATTER

Soil organic matter consists of a variety of components. *These include, in varying proportions and many Intermediate stages:*

- Raw plant residues and microorganisms (1 to 10 per cent)
- "Active" organic traction (10 to 40 per cent)
- Resistant or stable organic matter (40 to 60 per cent) also referred to as humus.

Raw plant residues, on the surface, help reduce surface wind speed and water run-off. Removal, incorporation or burning of residues predisposes the soil to serious erosion. The "active" and some of the resistant soil organic components, together with microorganisms (especially fungi) are involved in binding small soil particles into larger aggregates. Aggregation is important for good soil structure, aeration, water infiltration and resistance to erosion and crusting. The resistant or stable fraction of soil organic matter contributes mainly to nutrient holding capacity (cation exchange capacity) and soil colour. This fraction of organic matter decomposes very slowly and therefore has less influence on soil fertility than the "active" organic fraction. Organic matter in soil serves several functions. From a practical agricultural standpoint, it is important for two main reasons. First as a "revolving nutrient bank account"; and second, as an agent to improve soil structure, maintain tilth, and minimize erosion.

As a revolving nutrient bank account, organic matter serves two main functions:

1. Since soil organic matter is derived mainly from plant residues, it contains all of the essential plant nutrients. Accumulated organic

matter, therefore, is a storehouse of plant nutrients. Upon decomposition, the nutrients are released in a plant-available form.

2. The stable organic fraction (humus) adsorbs and holds nutrients in a plant available form.

Organic matter does not add any "new' plant nutrients but releases nutrients in a plant available form through the process of decomposition. In order to maintain this nutrient cycling system, the rate of addition from crop residues and manure must equal the rate of decomposition. If the rate of addition is less than the rate of decomposition, soil organic matter will decline and, conversely if the rate of addition is greater than the rate of decomposition, soil organic matter will increase. The term steady state has been used to describe a condition where the rate of addition is equal to the rate of decomposition. Fertilizer can contribute to the maintenance of this revolving nutrient bank account by increasing crop yields and consequently the amount of residues returned to the soil.

ORGANIC MATTER IN VIRGIN AND CULTIVATED SOILS

Soils in Alberta are divided into soil groups (zones) based on the amount of organic matter they contain. They occur in geographic zones from the southeast to the northwest and are identified as the Brown, Dark Brown, and Black Chernozemic (prairie) soils. The Brown soils have the least amount of organic matter because of the relatively small inputs of plant residues contributed by the short grass prairie vegetation under which these soils developed. Black soils developed under cooler and wetter conditions which allowed for more grass growth and thus a greater accumulation of organic matter.

Further north and west, trees became the dominant vegetation. Soils influenced by forest vegetation for a moderate length of time constitute the Dark Gray or transitional soils. Where the forest cover was established for a longer period, Luvisolic (forest) soils developed. Organic (peat) soils. occur in low lying areas throughout the Black, Dark Gray and Gray soil zones. These soils are saturated with water for much or all of the year thereby reducing the rate of organic matter decomposition. The amount of soil organic matter characteristic of virgin and cultivated soils in the various zones is shown in Table. Cultivation generally has resulted in a 30 to 50 per cent loss of organic matter.

Table. Organic Matter in Native and Cultivated Soils (per cent)

Soil zone	Virgin	Cultivated
Brown	3-4	2-3
Dark Brown	4-5	3-4
Black	6-10	4-6
Dark Gray	4-5	2-3
Gray	1-2	1-2

Before our soils were cultivated, they had achieved a "steady state". In most of our prairie soils, the increased rate of decomposition associated with cultivation, combined with the low rates of crop residue addition associated with crop-fallow rotations has caused a fairly rapid decline in soil organic matter. The rate of decline decreases with time as the amount of total soil organic matter decreases and particularly as the "active" organic fraction is depleted. Cultivation of soils that are naturally high in organic matter will usually result in a decrease of organic matter. In the case of Luvisolic soils, their poor physical properties and low fertility have encouraged the use of forages, fertilizers, manure and judicious tillage. Such management practices have resulted in an increase in soil organic matter on Luvisolic soils, whereas excessive tillage, fallowing and minimal fertilization have lead to further depletion of the soil organic matter.

EFFECTS OF ORGANIC MATTER DECLINE

As stated in the introduction, soil degradation is becoming a major concern in Canada. Loss of organic matter is often identified as one of the main factors contributing to declining soil productivity, but it is misleading to equate a loss in soil organic matter with a loss in soil productivity. Soil organic matter contributes to soil productivity in several ways, but there is no direct quantitative relationship between soil productivity and total soil organic matter. In fact, it has been the decline in organic matter that has contributed to the productivity of the crop-fallow system.

This decline in organic matter has resulted in the release of large amounts of plant nutrients, particularly nitrogen. For example, a decrease in soil organic matter of 2 per cent releases about 2,400 lb/ac of nitrogen. If this decline occurred over a 60 year period, an average of 40 lb/ac/yr of plant-available nitrogen has come from the soil organic matter. We therefore view prairie soils which had relatively high levels of organic matter as being nitrogen fertile, but this fertility could only be attained under a management system that allowed for organic matter to decline.

Frequent fallowing has been a major factor contributing to this decline. Insofar as organic matter contributes to improved soil physical properties (*e.g.*, tilth, aggregation, moisture holding capacity and resistance to erosion) increasing soil organic matter will generally result in increased soil productivity. But on many soils, suitable soil physical properties occur at relatively low levels of organic matter (2-4 per cent). A level of organic matter higher than required to produce suitable physical properties is beneficial in that the soil has a greater buffering and nutrient holding capacity, but it does not contribute directly to soil productivity. If soils are managed so organic matter is not declining (steady-state), soils higher in organic matter (*e.g.*, 8 per cent) are not inherently more productive or fertile than those that have less organic matter (*e.g.*, 5 per cent). To equate the ability to supply nutrients with total soil organic matter is not valid. The "active" fraction of organic matter is a more reliable indicator of soil fertility than is total soil organic matter. In cultivated soil, the "active" fraction is influence mainly by previous management. Soil organic matter cannot be increased quickly even when management practices that conserve soil organic matter are adopted. The increased addition of organic matter associated with continuous cropping, and the production of higher crop yields, are accompanied by an increase in the rate of decomposition. Moreover, only a small fraction of crop residues added to soil remains as soil organic matter. After an extended period of time, the return of all crop residues and the use of forages in rotations with cereals and oilseeds may significantly increase soil organic matter, particularly, the "active" fraction.

MANAGING SOIL ORGANIC MATTER

There have been vast changes in the nature of agricultural production. In the past, farms were small, and much of what was produced was consumed on the farm. This system allowed for the limited removal of soil nutrients since there was an opportunity to return most of the nutrients back to the land. The advent of the internal combustion engine, migration from rural to urban communities, increasing farm size and specialization in production have resulted in a system of production where there is greater removal of plant nutrients from the soil and less opportunity for nutrient cycling. Maintenance of organic matter for the sake of maintenance alone is not a practical approach to farming. It is more realistic to use a management system that will give sustained profitable production. The greatest source of soil organic matter is the residue contributed by current crops. Consequently, crop yield and type, method of handling residues and frequency of fallow are all important factors. Ultimately, soil organic matter must be maintained at a level necessary to maintain soil tilth.

Summerfallow

Summerfallowing accelerates the loss of organic matter. Aeration of the soil associated with tillage, and the increase in soil temperature and moisture results in increased organic matter decomposition. Since little In the way of residues are added to the soil, a net loss of organic matter occurs. Research has shown that as the frequency of fallow increases, the amount of soil organic matter decreases. Summer fallowing for moisture conservation may be a necessary practice in the Brown and Dark Brown soil zones. However, it must be questioned in the Black and the Gray soil zones. Periodic fallowing may be acceptable in the higher rainfall regions for control of persistent perennial weeds and volunteer grains in pedigreed seed production. In the crop-fallow system common to the prairie region, the nitrogen removed has far exceeded that gained from crop residues, manure, legumes and fertilizer.

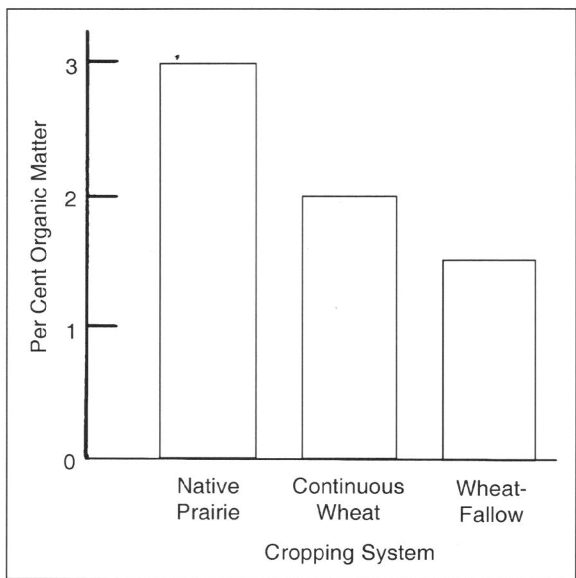

Fig. Effect of Frequency of Fallow on Per Cent Organic Matter.

The large reserves of nitrogen present in the organic matter of our prairie soils have been the major source of nitrogen in this cropping system. Continued reliance on soil organic matter reserves to supply the nitrogen requirements of crops will ultimately lead to a decline in soil productivity, and increased soil erosion. When a change from a cropping system involving fallow to continuous cereal grain production, the nitrogen requirement increases. The nitrogen requirement is greatest in the first few years of continuous cropping as the nutrient cycling process adjusts to the new cropping system.

Crop Rotations

The value of forage crops in rotations with cereals and oilseeds has long been recognized, especially in the Luvisolic soils. The Breton Plots compared a two-year fallow-wheat rotation with a five-year rotation involving wheat, oats and barley followed by two years of hay production. In research done by the University of Manitoba, the effects of various cultural practices on the level of organic matter are compared.

It is interesting to note that the highest level of soil organic matter was maintained under continuous cropping.

The beneficial effects of perennial forages are the result of:

- A more extensive root system and crop aftermath contributing more organic matter to the soil,
- The fibrous nature of the root system of perennial grasses. These are particularly effective as a binding agent in soil aggregation,
- Nitrogen fertility enhancement by the growth of legumes,

- Increased permeability of dense subsoils because of the deep penetrating tap roots of perennial legumes, especially alfalfa.
- A reduced rate of organic matter decomposition in the absence of tillage.

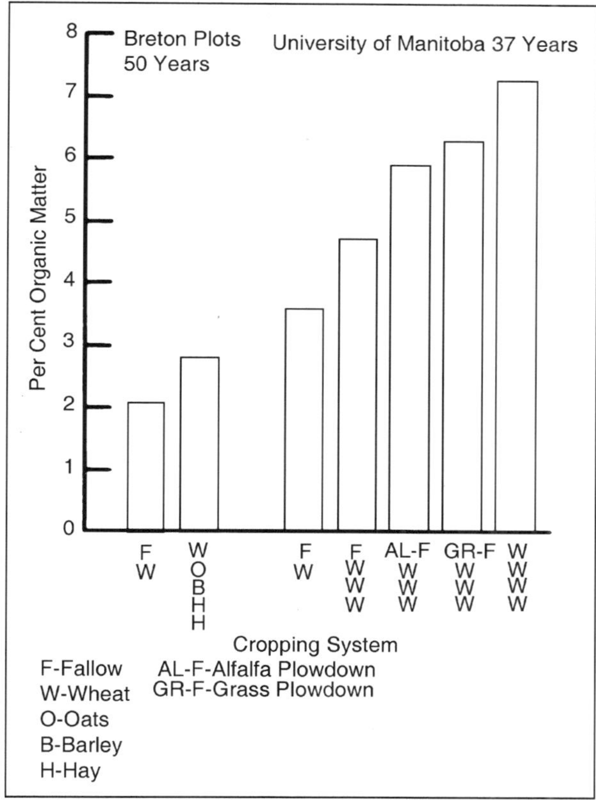

Fig. Effect of Rotation on Soil Organic Matter

In the brown and dark brown soil zones, perennial forages grown for forage production or as a plowdown crop may jeopardize subsequent cereal crops because they deplete soil moisture reserves.

Fertilization

Fertilizers will generally increase soil organic matter because the increased crop growth returns larger amounts of residues to the soil. Data obtained from the Breton Plots is summarized in Figure. The increase in organic matter is less than what might be expected with current farming practices since all the straw had been removed from the plots.

To determine the fertilizer effect, two fertilizer treatments were averaged, one involving a low rate of nitrogen and sulphur and another involving a low rate of nitrogen, phosphorus, potassium and sulphur.

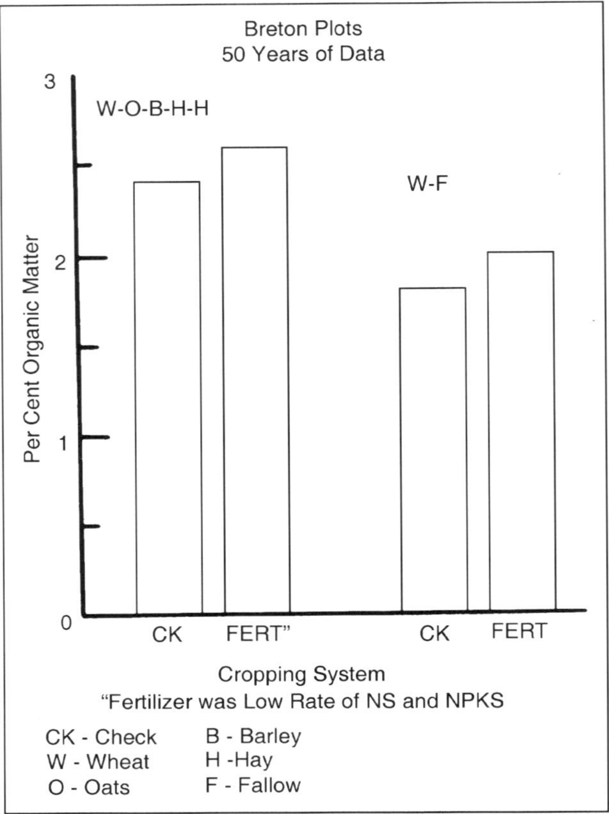

Fig. Effect of Fertilizer on Soil Organic Matter.

Plowdown

Legume plowdown has received considerable attention in recent years as an alternative to the use of nitrogen fertilizers. However, when considering this option in a cropping programme, the amount of nitrogen added by the legume, as well as the loss of one year of production, the cost of seed and the expected yield increase must be kept in mind. Strictly as a source of nitrogen, the value of a legume plowdown is questionable.

The amount of nitrogen fixed by a legume is dependent upon the type of legume, the amount of vegetative growth, the nature of the soil and environmental conditions.

As a source of organic matter, legume plowdown is valuable, however, perennial forage is more effective than legume plowdown for increasing soil organic matter. Nitrate nitrogen which accumulates following legume plowdown is subject to loss, particularly in wet, poorly drained soils. To minimize this, legumes should be plowed down in the fall rather than mid-summer to reduce nitrate accumulation and subsequent loss.

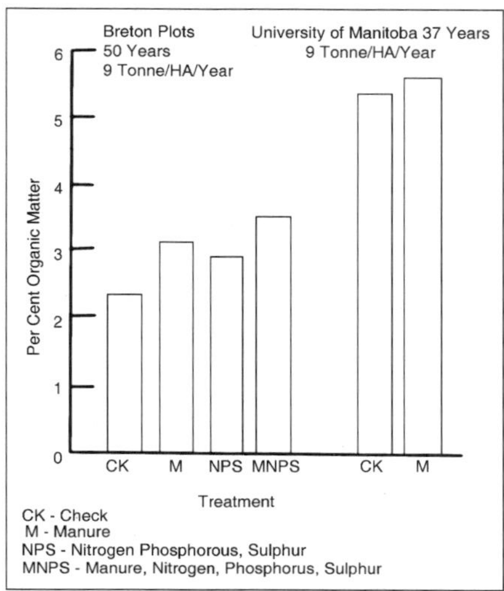

Fig. Effect of Manure on Soild Organic Matter.

The cultivation of prairie soils has generally resulted in a decline in organic matter of 30 to 50 per cent. A product of this decline has been the release of large amounts of plant nutrients, particularly nitrogen. Crop rotations with a high frequency of summerfallow have relied on the nitrogen released from soil organic matter to supply crop requirements. More frequent or continuous cropping, less frequent tillage, the production of high yields and the return of crop residues will help to maintain soil organic matter at a satisfactory level. Perennial forages are effective for maintaining or increasing soil organic matter.

SOIL MICROBIAL ACTIVITY

A rapid increase in soil microorganisms occurs after a young, relatively lush green manure crop is incorporated into the soil. The soil microbes multiply to attack the freshly incorporated plant material. During microbial breakdown, nutrients held within the plant tissues are released and made available to the following crop. Factors that influence the ability of microorganisms to break down organic matter include soil temperature, soil moisture, and carbon to nitrogen (C:N) ratio of the plant material. The C:N ratio of plant tissue reflects the kind and age of the plants from which it was derived.

C:N ratios above 25:1 can result in nitrogen being "tied up" by soil microbes in the breakdown of carbon-rich crop residues, thus pulling nitrogen away from crop plants. Adding some nitrogen fertilizer to aid the decomposition process may be advisable with these high carbon residues. The lower the C:N ratio, the more N will be released into the soil for immediate crop use. The C:N ratio is more a function of the plant's N content than its carbon content. Most plant materials contain close to 40 per cent carbon. To determine the C:N ratio of any plant material, divide 40 per cent by its nitrogen content. For example let's say hairy vetch contains 4.2 per cent nitrogen: 40/4.2= a C:N ratio of 9.5. A procedure for determining the nitrogen content of cover crop biomass was previously addressed in the section on nitrogen production. Estimating the nitrogen contribution of a cover crop is very helpful when adjusting N-fertilizer rates to account for legume nitrogen.

NUTRIENT ENHANCEMENT

In addition to nitrogen from legumes, cover crops help recycle other nutrients on the farm. Nitrogen (N), phosphorous (P), potassium (K), calcium (Ca), magnesium (Mg), sulfur (S), and other nutrients are accumulated by cover crops during a growing season. When the green manure is incorporated, or laid down as no-till mulch, these plant-essential nutrients become slowly available during decomposition. Dr. Greg Hoyt developed a method for estimating nutrient accruement by cover crops in order to reduce the soil test recommendation of fertilizer for the following crop.

Rooting Action

The extensive root systems of some cover crops are highly effective in loosening and aerating the soil. In Australian wheat experiments, the taproots of a blue lupine cover crop performed like a "biological plow" in penetrating compacted soils. When cover crops are planted after a subsoiling treatment, they help extend the soil-loosening effects of the deep tillage treatment. The rooting depths of several green manures grown under average conditions are listed in.

Weed Suppression

Weeds flourish on bare soil. Cover crops take up space and light, thereby shading the soil and reducing the opportunity for weeds to establish themselves. The soil-loosening effect of deep-rooting green manures also reduces weed populations that thrive in compacted soils. The primary purpose of a non-legume green manure—such as rye, millet, or sudangrass—is to provide weed control, add organic matter, and improve soil tilth. They do not produce nitrogen. Thus, whenever possible, annual grain or vegetable crops should follow a legume green manure to derive the benefit of farm-produced nitrogen.

Providing weed suppression through the use of allelopathic cover crops and living mulches has become an important method of weed control in sustainable agriculture. Allelopathic plants are those that inhibit or slow the growth of other nearby plants by releasing natural toxins, or "allelochemicals." Cover crop plants that exhibit allelopathy include the small grains like rye and summer annual forages related to sorghum and sudangrass. The mulch that results from mowing or chemically killing allelopathic cover crops can provide significant weed control in no-till cropping systems. Living mulches suppress weeds during the growing season by competing with them for light, moisture, and nutrients.

Soil and Water Conservation

When cover crops are planted solely for soil conservation, they should provide a high percentage of ground coverage as quickly as possible. Most grassy and non-legume cover crops, like buckwheat and rye, fulfil this need well. Of the winter legumes, hairy vetch provides the least autumn ground cover because it puts on most of its above-ground growth in the spring. Consequently, it offers little ground cover during the erosion-prone fall and winter period. Sowing a mix of leguminous and grassy-type cover crops will increase the ground coverage, as well as provide some nitrogen to the following crop. The soil conservation benefits provided by a cover crop extend beyond protection of bare soil during non-crop periods. The mulch that results from a chemically or mechanically killed cover crop in no-till plantings increases water infiltration and reduces water evaporation from the soil surface. Soil cover reduces soil crusting and subsequent surface water run-off during rainy periods.

Retention of soil moisture under cover crop mulches can be a significant advantage. Dr. Blevins and other researchers showed consistently higher soil-moisture levels for corn grown in a herbicide-killed, no-till bluegrass sod than for corn grown in conventionally plowed and disked plots. They concluded that the decreased evaporation and increased moisture storage under the no-till mulch allowed plots to survive a short-term drought without severe moisture stress.

Vegetation Management to Create a Cover Crop Mulch

Herbicides are the most commonly used tools for cover crop suppression in conservation tillage systems. Non-chemical methods include propane flamers, mowing and mechanical tillage. Mowing a rye cover crop when it heads out in late spring provides sufficient kill. The rye must be in the pollination phase, or later, to be successfully killed. When the anthers are fully extended and you can thump the stalk and pollen falls down, it is time to mow. If mowed earlier, it just grows back. Flail mowers generally produce more uniformly distributed mulch than do rotary cutters, which tend to windrow the mulch to one side of the mower. Sickle bar mowers create fairly uniform mulch, but the unchopped rye stalks can be more difficult to plant into. If late spring weather continues cool and wet, more rye regrowth will occur than if the weather remains warm and drier. Typically, if rye is mowed at the pollination stage, regrowth is minimal and not a problem to crops grown in the mowed mulch.

In a Mississippi study, flail mowing, or rolling with rolling disk colters spaced at 4 inches, was usually as effective as herbicides in killing hairy vetch, crimson clover and subterranean clover. Timing is a key factor when using mowing or rolling to control cover crops. Mechanical control was most effective when the legumes were in the seed formation growth phase (mid to late April) or when stem lengths along the ground exceeded 10 inches. If mowing was followed with a pre-plant herbicide application of Atrazine, the legume kill was even more effective.

Researchers at Ohio State University developed a mechanical cover crop killing tool used to take out a cover crop without herbicides. They call it an undercutter because it uses wide V-blades which run just under the soil surface to cut off the cover crop from its roots. The blades are pitched to 15 degrees, allowing the blades to penetrate the soil and provide a slight lifting action. Mounted on the same toolbar behind the cutter blades is a rolling basket to flatten and distribute the undercut cover crop. The undercutter was tried on several cover crops and effectively killed crimson clover, hairy vetch, rye, and barley. These undercutters could be made from locally available stock by innovative tinkerers. Steve Groff of Cedar Meadow Farm in Lancaster County, Pennsylvania, uses a 10-foot Buffalo rolling stalk chopper from Fleischer Manufacturing. Under the hitch-mounted frame, the stalk chopper has two sets of rollers running in tandem.

These rollers can be adjusted for light or aggressive action and set for continuous coverage. Steve says the machine can be run up to 8 miles an hour and does a good job of killing the cover crop and pushing it right down on the soil. It can also be used to flatten down other crop residues after harvest. Groff improved his chopper by adding independent linkages and springs to each roller. This modification makes each unit more flexible to allow continuous use over uneven terrain. Following his chopper, Groff transplants vegetable seedlings into the killed mulch. He direct-seeds sweet corn and snap beans into the mulch. Two USDA-ARS researchers, Drs. Aref Abdule-Bake and John Teasdale of the Beltsville Maryland Research Centre, have developed a cover-crop roller that acts, in principle, similarly to Steve Groff's rolling chopper.

In their extensive research trials using hairy vetch, they no-till planted tomatoes into a mechanically killed hairy vetch cover crop. Details of their research—and other useful information on flail-mowing of cover crops and direct no-till seeding of sweet corn and snap beans into mechanically killed cover crops—can be seen in the USDA Farmer's Bulletin No. 2279, listed under the Web Resources section below. As of this writing the bulletin is available only online because the first printing of it was all distributed to farmers and their advisors in a very short time.

Planting cover crops known to readily winter-kill is another non-chemical means of vegetation management. Spring oats, buckwheat, and sorghum fill this need. They are fall-planted early enough to accumulate some top growth before freezing temperatures kill them. In some locations, oats will not be completely killed and some plants will regrow in the spring. Winter-killed cover crops provide a dead mulch through the winter months instead of green cover. They are used primarily in regions where precipitation is limited, such as West Texas, and in situations where farmers want to plant early in the spring and avoid over wintered cover crops altogether.

LIMITATIONS OF COVER CROPS

The recognised benefits of green manuring and cover cropping—soil cover, improved soil structure, nitrogen from legumes—need to be evaluated in terms of cash returns to the farm as well as the long-term value of sustained soil health. For the immediate growing season, seed and establishment costs need to be weighed against reduced nitrogen fertilizer requirements and the effect on cash crop yields.

Water consumption by green manure crops is a concern and is pronounced in areas with less than 30 inches of precipitation per year. Still, even in the fallow regions of the Great Plains and Pacific Northwest, several native and adapted legumes (such as black medic) seem to have potential for replacing cultivation or herbicides in summer fallow. There is always additional management required when cover crops of any sort are added to a rotation. Turning green manures under or suppressing cover crops requires additional time and expense, compared to having no cover crop at all.

Insect communities associated with cover crops work to the farmer's advantage in some crops and create a disadvantage in others. For example, certain living mulches enhance the biological control of insect pests of summer vegetable crops and pecan orchards by providing favourable habitats for beneficial insects. On the negative side, winter legumes that harbour catfacing insects such as the tarnished plant bug, stink bug, and plum curculio can pose problems for apple or peach orchardists in the eastern U.S. Nematodes encouraged by certain legumes on sandy soils are another concern of farmers, as are cutworms in rotations following grain or grass crops.

Cover Crops in Rotation

Cover crops can fit well into many different cropping systems during periods of the year when no cash crop is being grown. Even the simplest corn/ soybean rotation can accommodate a rye cover crop following corn, which will scavenge residual nitrogen and provide ground cover in the fall and winter. When spring-killed as a no-till mulch, the rye provides a water-conserving mulch and suppresses early-season weeds for the following soybean crop. Hairy vetch can be planted behind soybeans to provide nitrogen for corn the following spring. Hairy vetch is *not* a good cover crop to use when small grains are included in the rotation—if the vetch ever goes to seed it can become a terrible weed in the small grain crop. In these cases, crimson clover, sweet clover, or red clover should be used, depending on location.

Many vegetable rotations can accommodate cover crops as well. Buckwheat can follow lettuce and still be tilled down in time for fall broccoli. Hairy vetch works well with tomatoes and other warm-season vegetables. The vetch can be killed by flail mowing and tomato sets planted into the mulch.

Pest Management Benefits of Cover Crops

In addition to the soil improving benefits, cover crops can also enhance many pest management programmes. Ecologists tell us that stable natural systems are typically diverse, containing many different types of plants, arthropods, mammals, birds, and microorganisms. Growing cover crops adds diversity to a cropping system. In stable systems, serious pest outbreaks are rare, because natural controls exist to automatically bring populations back into balance.

Farmers and researchers in several locations have observed and documented increased beneficial insect numbers associated with cover crops. The cover crops provide pollen, nectar, and a physical location for beneficial insects to live while they search for pest insects. Conservation tillage proves a better option than tilling because it leaves more crop residue on the surface to harbour the beneficial insects. Strip tilling or no-tillage disturbs a minimum of the existing cover crop that harbours beneficial insects. Cover crops left on the surface may be living or in the process of dying. At either of these stages they protect beneficials. Once the main crop is growing, the beneficials move onto it. By having the cover crop in place early in the growing season, the population of beneficials is much higher sooner in the growing season than would be the case if only the main crop were serving as habitat for the beneficials.

Innovative farmers are paving the way by interplanting cover crops with the main crop and realising pest management benefits as a result. Georgia cotton farmers Wayne Parramore and sons reduced their insecticide and fertilizer use by growing a lupine cover crop ahead of their spring-planted cotton. They started experimenting with lupines on 100 acres in 1993 and by 1995 were growing 1,100 acres of lupines. Ground preparation for cotton planting is begun about 10 days prior to planting by tilling 14-inch wide strips into the lupines. Herbicides are applied to the strips at that time and row middles remain untouched. The remaining lupines provide beneficial insect habitat and also serve as a smother crop to curtail weeds and grasses. The lupines in the row middles can be tilled in later in the season to release more legume nitrogen.

Dr. Sharad Phatak of the University of Georgia has been working with cotton growers in Georgia testing a strip cropping method using winter annual cover crops. Planting cotton into strip-killed crimson clover improves soil health, cuts tillage costs, and allows him to grow cotton without any insecticides and only 30 pounds of nitrogen fertilizer. Working with Phatak, farmer Benny Johnson reportedly saved at least $120/acre on his 16-acre test plot with the clover system.

There were no insect problems in the test plot, while beet armyworms and whiteflies infested nearby cotton and required 8 to 12 sprays to control. Cotton intercropped with crimson clover yielded more than 3 bales per acre compared with 1.2 bales per acre in the rest of the field. Boll counts were 30 per plant with crimson clover and 11 without it. Phatak identified up to 15 different kinds of beneficial insects in these strip-planted plots. Phatak finds that planting crimson clover seed at 15 pounds per acre in the fall produces around 60 pounds of nitrogen per acre by spring. By late spring, beneficial insects are active in the clover. At that time, 6- to 12-inch planting strips of clover are killed with Roundup herbicide. Fifteen to 20 days later the strips are lightly tilled and cotton is planted. The clover in the row-middles is left growing to maintain beneficial insect habitat.

When the clover is past the bloom stage and less desirable as a preferred habitat, beneficials move onto the cotton. Even early-season thrips, which can be a problem following cover crops, are limited or prevented by beneficial insects in this system. Movement of the beneficials coincides with a period when cotton is most vulnerable to insect pests. Following cotton defoliation in the early fall, the beneficials hibernate in adjacent non-crop areas.

Phatak points out that switching to a whole-farm focus while reducing off-farm inputs is not simple. It requires planning, management, and several years to implement on a large scale. It is likewise important to increase and maintain organic matter, which stimulates beneficial soil microorganisms. Eventually a "living soil" will help keep harmful nematodes and soil-borne fungi under control.

Economics of Cover Crops

The most obvious direct economic benefit derived from legume cover crops is nitrogen fertilizer savings. In most cases these savings can offset cover crop establishment costs. Indirect benefits include herbicide reduction in the case of an allelopathic rye cover crop, reduction in insect and nematode control costs in some cases, protection of ground water by scavenging residual nitrate, and water conservation derived from a no-till mulch. Longer-term benefits are derived from the buildup of organic matter resulting in increased soil health. Healthy soils cycle nutrients better, don't erode, quickly absorb water after each rain, and produce healthy crops and bountiful yields.

With annual cover crops, the highest cost is seed. Hairy vetch and crimson clover typically range from 50¢ to $1.50 per pound. With a 20-pound per acre seeding rate, seed costs range from $10 to $30 per acre. With a 25-pound seeding rate at 85¢/lb and a $6.50 no-till drilling cost, it would cost $28 to plant an acre of this cover crop. In a Maryland study, hairy vetch was compared to a winter wheat cover crop or no cover crop at two different locations (coastal plain and piedmont). Corn was grown following the cover crops. Nitrogen fertilizer was used with the cover crops at varying rates. The most profitable cover crop and nitrogen fertilizer combination used more than 100 lbs of additional nitrogen per acre plus the cover crop. At $2.50 per bushel corn price, highest returns at the coastal plain location were realised with 120 lbs of additional nitrogen per acre. Profits were as follows: $53.75 per acre from no cover crop, $95.62 from hairy vetch, and $32.47 from winter wheat cover crop. All corn crops needed additional nitrogen. Lower N rates were less profitable. At the piedmont location, also with $2.50 corn, winter fallow was most profitable at $68.03 with 40 lbs per acre additional N, hairy vetch was profitable at $56.57 with 40 lbs per acre, and winter wheat was profitable at $30.12 with 100 lbs of additional nitrogen.

In another Maryland study, optimum nitrogen rates for corn were determined when corn followed four cover crops, compared to a winter fallow (no cover crop) treatment. Corn was grown following each cover crop treatment at various nitrogen rates over a three-year period. The optimum nitrogen rate is the rate above which no additional yield increases are realised. The researchers concluded that cover crops can benefit a succeeding corn crop not only by supplying nitrogen but also by increasing maximum yield of the system.

ORGANIC NITROGENOUS FERTILIZERS

These fertilizers include plant and animal by-products, such as oil cakes, fish manure and dried blood from slaughter houses. Before their organic nitrogen can be used by the crops, it is converted through bacterial action into readily usable ammonia-nitrogen and nitrate nitrogen. These fertilizers are, therefore, relatively slow-acting, but they supply available nitrogen for a longer periods. Furthermore, they may also small amounts of organic stimulants that they may contain, or of some of the minor elements needed by plant.

Oil-cakes of different kinds are produced in India to the tune of about two million tonnes annually. They contain not only nitrogen but also some phosphoric and potash, besides a large quantity of organic matter. The chemical composition of the principal oil-cakes available in the country is given in Table. In addition to the three fertilizing constituents (N, P_2O_5 and K_2O), the oil-cakes invariably contain 2 to 15 per cent of oil, depending on whether the oil is extracted by using solvent process or with expellers, hydraulic presses or indigenous ghanis.

This residual oil, however, dous not, in practice, affect their manurial value. Owing to the great importance of using edible oil-cakes as a cattle feed, their utilization as fertilizers is undesirable. They should be fed to cattle and the excrements may be used as manure. Inedible cakes, like castor cake, neem, mahua cake and karanj cake can, however, be recommended for use in conjunction with quicker-acting chemical fertilizers.

Dried blood or blood-meal contains 10 to 12 per cent highly available nitrogen and 1 to 2 per cent phosphoric acid. It is a very quick-acting manure and effective on all crops and all types of soils. It should be used in the same way as oil cakes. Fish manure is available either as dried fish or as fish-meal or powder. In regions where fish oil is extracted, the residue can be used as a manure. Depending on the type of fish, its manurial constituents vary from 5 to 8 per cent of organic nitrogen and from 4 to 6 per cent of phosphoric acid. It is quick-acting and suitable for all crops and soils. It should preferably be powdered before use.

PHOSPHATE FERTILIZERS

These are classified as natural phosphates, treated or processed phosphates, and by-product phosphates and chemical phosphates.

Rock Phosphate

It occurs as natural deposits of rock in Morocco, the United States of America, Poland, Russia, Tunisia, Algeria, Algeria, Brazil, Egypt, Nehru and some islands in the Pacific Ocean and the Indian Ocean. It contains 25 to 35 per cent phosphoric acid, but this phosphorus is insoluble in water. Practically no rock phosphate, as such, is used as a fertilizer in this country, except some quantity in southern India. In other countries, finely pulverized phosphate rock has been found to give satisfactory results in soils which are very deficient in phosphorous and are acidic. Adequate rainfall and a long growing period of the crop enhance the response. All the same, very little rock phosphate is used is directly as a fertilizer even in these countries. Much more of it is used to manufacture superphosphate, the phosphoric acid of which is water-soluble and is in an available form.

Superphosphate

It is the most widely used fertilizer in India. Prepared formerly by treating bones with sulphuric acid, it is now manufactured largely by treating ground phosphate rock with almost an equal quantity by weight of sulphuric acid. This treatment produces a brownish-grey mixture containing monocalcium phosphate and calcium sulphate (gypsum) in practically equal quantities. This fertilizer is manufactured in three grades: single superphosphate containing 16 to 20 per cent phosphoric acid; dicalcium phosphate, 35 to 38 per cent; and triple superphosphate, 44 to 49 per cent. Single superphosphate is the most commonly available grade in the Indian market. Triple superphosphate in the production of which liquid phosphoric acid is used instead of sulphuric acid, contains very little calcium sulphate. Tripple superphosphate is used mostly in the manufacture of concentrated mixed fertilizers. The phosphoric acid in superphosphate is wholly water-soluble but when applied to the soil, it is immediately converted into insoluble phosphate owing to precipitation as calcium, iron or aluminium phosphate, according as the soil is alkaline or acid. Thus the fertilizer is not leached, but is slowly dissolved in the soil solution. Fixation losses can be reduced by applying the fertilizer in bands on both sides of the row of seeds at a depth of 10 to 15 cm with a drill. By this means, at least some of the phosphate does not come into direct contact with the soil and thus the available phosphorous is readily released for absorption by the plant roots.

The fertilizer is suitable for all crops and can be applied to all soils. In acid soils, it should properly be used in conjunction with organic manure. It should be applied before or at sowing or transplanting.

Basic-slag

This is a by-product of steel factories. Depending on the phosphorous content of the iron ore, it contains from 6 to 20 per cent of phosphoric acid (P_2O_5). Slag from Indian steel-mills is poor in P_2O_5 and is not used as a fertilizer. The high-grade European slag containing 15 to 18 per cent P_2O_5 is a popular phosphatic fertilizer in central Europe. It is not as soluble as superphosphate, but unlike the latter, it is alkaline in reaction and good for acid soils. For effective use, it must be pulverized before application.

Bone-meal

The use of raw bones as a manure for fruit-trees is an age-old practice in this country. Burying the skeleton of an animal under a fruit-tree is known to benefit its growth and bearing. Large quantities of bones used to be exported until a few years ago. Exports have since, declined and bone-meal (*i.e.,*, ground bone) is now a widely used phosphate fertilizer.

It is available in two forms:

1. Raw bone-meal,
2. Steamed bone-meal.

The steaming of bones under pressure removes fats, greases, nitrogen and glue-making substances. Thus, while raw bone-meal contains about 4 per cent slow-acting organic nitrogen and 20 to 25 per cent insoluble phosphoric acid.

Steamed bones are more brittle and can be readily ground. This is an advantage, as the rate of availability of phosphoric acid. Steamed bones are more brittle and can be readily ground. This is an advantage, as the rate of availability of phosphoric acid in bones depends largely on their degree of pulverization. Bone-meal contains only 1 to 2 per cent nitrogen but 25 to 30 per cent phosphoric acid. Steamed bones are more brittle and can be readily ground. This is an advantage, as the rate of availability of phosphoric acid in bones depends largely on their degree of pulverization. Bone-meal, having particles not larger than 3/32 inch, considered suitable for use as a fertilizer, but the more finely powdered it is, the quicker its P_2O_5 becomes available in the soil. Being relatively slow-acting, bone-meal should not be used as a top-dressing; it must be incorporated into the soil in order to become available. It may be applied either at sowing time or a few days before sowing and should be broadcast. It is particularly suitable for acid soils. It is considered a safe manure for all crops.

In some parts of the country, charred and powdered bones are used as a manure. Charring destroys about half the nitrogen, but leaves intact practically the whole of P_2O_5 in a quickly available form. In the absence of arrangements for steaming and grinding, charring can be easily carried out even in remote villages.

Potassic Fertilizers

Most of the Indian fertilizers soils contain a sufficient amount of potash. Potassic fertilizers should, therefore, be applied only to such soils as are definitely known to be deficient in potash or to those which respond to their application, such as sandy soils. They can also be applied to certain crops, such as tobacco, potato, onion, tomato and fruit-trees, to improve the quality and appearance of their produce.

Potassic fertilizers in common use are:

- Muriate of potash (potassium chloride), and
- Sulphate of potash (potassium sulphate).

These salts are important constituents of the waters of oceans and inland seas and of saline deposits derived thereform. The largest known deposits of these salts are at Stassfurt in Germany, in the Caspian Sea region in Russia, in the Dead Sea in Palestine and at some places in California, New Mexico, France and Spain.

Muriate of Potash

It is a gray crystalline material containing 50 to 63 per cent potash (K_2O), the whole of which is readily available. Though highly soluble in water, it is not lost from the soil, as it is absorbed on the colloidal surfaces. It can be applied at sowing time or before sowing.

Sulphate of Potash

It is made by treating potassium chloride with magnesium sulphate and is, therefore, more costly. It contains 48 to 52 per cent K_2O. It dissolves readily in water and becomes available to the crop almost immediately. It can be applied at any time up to sowing, but should not be drilled with the seed. It is considered better than muriate of potash for crops, such as tobacco, chillies, potato and fruit-tree, where quality is of prime importance.

Other Sources

Wood ashes, cattle-dung ash, leaf-mould, tobacco stems and water hyacinth are the available indigenous sources of potash. Unleached wood ash contains 5 to 6 per cent of potash in the form of potassium carbonate (which is alkaline), 1 to 2 per cent phosphoric acid and 25 to 30 per cent lime (Cao). Both potassium carbonate and lime in the wood ashes counteract the acidity in the soil. Groundnut shell, paddy husk and bagasse ashes are available near decorticating factories and rice and sugar-mills.

These also contain a fair amount of potash and some phosphoric acid. Ground tobacco stems contain 2 to 3 per cent of nitrogen and 6 to 10 per cent of potash in quickly available forms.

Hyacinth abounds as a weed in freshwater ponds in Bengal, Assam, Tripura and Malabar in Kerala. When dry, it contains 1 per cent nitrogen, 4 per cent potassium and a small quantity of phosphorous.

Soil Amendments

Lime is generally used for correcting soil acidity, for improving the physical condition of the soil and encouraging bacterial activity. Similarly, gypsum is used for reclaiming 'alkali' soils or land from the sea and improving the structure of heavy black clay soils. Hence, these are called soil amendments.

Compound Fertilizers.

These fertilizers are multiple nutrient materials, supplying two or three plant nutrients simultaneously. When both nitrogen and phosphorous are deficient in a soil, a compound fertilizer, *e.g.*, ammophos, can be used. It contains 16 per cent N and 20 per cent P_2O_5.

Its use does away with the necessity of purchasing two different fertilizers and mixing them in correct proportion before use. The nutrient contents of some of the other compound fertilizers are shown below:

	N	**P_2O_5**	**K_2O**
Monoammonium phosphate	11.0	48.0	-
Diammonium phosphate Ammoniated superphosphate(4 - 7 per cent)	2.0 - 4.0	14.0 - 20.0	-
Potassium nitrate	13.0	-	44.0

Mixed Fertilizers

Compound fertilizers contain plant food elements in fixed proportions and are, therefore, not always best adapted in different kinds of soils. Accordingly, the needs of different soils can generally be met most economically by the use of fertilizer mixtures containing two or more materials in suitable proportions. Mixtures usually meet nutrient deficiencies in a more balanced manner and require less labour to apply than straight fertilizers used separately. Mixtures containing all the three principal used separately. Mixtures containing all the three principal nutrients (N, P and K) are termed complete fertilizers.

Some manufacturers prepare special mixtures for different crops, such as wheat, sugarcane, pady, potatoes, tabacco, fruit-trees and vegetables. Trade names like Ammophos, Niciphos and Nitrochalk are employed for other proprietary products, In foreign countries, even insecticides, fungicides and weed-killers, such as DDT, BHC and mercury or copper salts and 2, 4-D are sometimes incorporated into fertilizers mixtures.

When a required mixture is not available of the cost of the mixture sold in the market is high in relation to the cost of its individual components, it may be made at home by mixing the constituent fertilizers in correct proportion. The preparation of mixed fertilizers requires a good knowledge of the properties and mutual reaction of the component straight fertilizers under different climatic and storage conditions. Therefore in preparing such mixtures at the farm or at home, care should be taken to avoid the uneven mixing of incompatible fertilizers, or the mixing which leads to a loss of some of the fertilizing nutrients in the form of gas, converts soluble nutrients into insoluble ones or induces caking.

It is unwise to mix the following:

- Ammonium sulphate, ammonium chloride, other ammoniacal fertilizers and nitrogenous organic manures with lime.
- Sodium nitrate or potassium nitrate with superphosphate.
- Nitrochalk with superphosphate or lime.
- Ammonium sulphate-nitrate with lime.
- Urea with superphosphate.
- Superphosphate with lime or calcium carbonate or wood ashes.

Ammonium nitrate is an explosive chemical and, therefore, dangerous should be mixing at home. Mixtures containing other nitrates should be made in only such quantities as are to be used immediately, because they absorb water quickly and are not suitable for storage. Bone meal, sulphate of potash and muriate of potash can be mixed with all fertilizers. For further guidance or information on the method of mixing fertilizers, the State Agricultural Chemist or the nearest officer of the Government Agricultural Department should be contacted.

Mode and Time of Application of Fertilizers and Manures

Bulky organic manures should be applied well ahead of sowing, so that the preliminary decomposition takes place before the seeds germinate. Failing presowing application, they may be applied any time after the seedlings have established themselves. They are best applied in the powdered form. A sufficient supply of moisture in the soil is essential for their rapid decomposition. In the case of inorganic fertilizers, potassic and phosphatic fertilizers are best applied just before sowing or transplanting. Nitrogenous fertilizers may be applied either at planting and partly later. Split application is particularly desirable for nitrogen when applied to irrigated crops or to crops in heavy-rainfall areas.

Fertilizers applied before sowing should be broadcast uniformly and harrowed in. In the case of fertilizers containing soluble phosphate, the desirability of applying them in 2.5 to 5 cm wide bands on each side of the row of seeds at a depth of 10 to 15 cm with a drill has already been pointed out. This operation reduces the fixation of soluble phosphate in the soil. It is also a good practice to mix superphosphate with farmyard manure at 18 to 22 kg to a tonne before applying the organic manure, particularly of dairy farms land. Sulphate of ammonia used as a top-dressing should not be applied when plant leaves are wet. In the case of irrigated crops, the application of fertilizer should invariably be followed by a watering. In the case of fruit-trees, the fertilizer should be applied to the soil under the crown, a few metres away from the trunk. The area of application should be progressively extended as the trees grow bigger. In advanced countries, fertilizers are usually applied with the help of machinery of diverse kinds and sometimes with an aeroplane or a helicopter. Combined-planters and fertilizer-distributors are employed when row crops are fertilized at sowing time.

SIGNIFICANCE OF SOIL STRUCTURE

Soil structure also determines the amount and arrangement of empty spaces in the soil, and thus has great influence on how readily water moves through the soil and where plant roots can grow. When the soil is saturated (holds all the water it can), water moves down the cracks and pores in the soil. Roots also penetrate through these open spaces. Shape of the peds. The path water and roots must take to get through the soil. Granular and blocky peds are about the same size in all directions. In prismatic peds, the long direction is vertical and the cracks between peds are mainly vertical, the same direction that roots and water move.

In platy structure, however, the long direction is horizontal so roots or water first move downward between peds then must move horizontally across the plates before they find another vertical space. This complex path retards movement of water and growth of roots. Size of the peds determines the distance between the natural cracks. For example, soils such as Brookston have a prismatic structure in which many of the prism-shaped peds are about 1/2-1 inch across. Fragipan soils such as Avonburg, on the other hand, have prisms more than 4 inches across. Thus, Brookston with its smaller prisms has many more cracks through which roots can penetrate and water can flow. Strength of development is yet another aspect of soil structure. In strongly-developed structure, the peds are quite obvious, and the soil breaks up readily into peds when dug into or picked with a knife or spatula. Often soil peds have coatings, such as clay films. In weakly-developed structure, the peds are less apparent. Strength of structure development is very important in surface horizons. When raindrops hit a horizon with weakly developed structure, such as the gray surface layer of poorly drained soils on flat landscape positions (Clermont and Cobbsfork soils), the structure is destroyed. Individual soil particles are released and they plug up soil pores and prevent water from entering the soil. When the soil dries, it has a strong surface crust that may prevent seedlings from emerging. A soil with a strong structure, such as the dark surface layer of a prairie soil (Parr), will better withstand raindrop impact and not seal and crust so much.

O Horizon

This horizon consists of organic materials on the surface of the mineral soil such as in the Trappist soil (Plate 10). In this soil, the organic materials are predominantly decomposing leaves from hardwood trees. In bogs and marshes, the whole soil formed in organic matter, and the entire profile consists of O horizons.

A Horizon

This is the mineral horizon at the surface. It is the one in which living organisms are most active and, therefore, is marked by the dark coloured humus mixed with the mineral material Under forest vegetation, the A horizon is only a few inches thick (Miami, Plate 2). But under prairie grasses (Parr, Plate 5) or in swales (Brookston, Plate 4), the A may be 10-25 or more inches thick. The A usually has the darkest colour of any horizon in the soil. The Ap horizon includes the part of the soil that has been mixed by plowing or cultivation. It consists mostly of the A horizon, the A and E or, in eroded soils, the remaining A and part of the upper B horizons. Soils developed under grass vegetation usually have an AB horizon, transitional between the A and B horizons.

E Horizon

Forest soils commonly contain E horizons a few inches below the surface. An E is lighter in colour, lower in organic matter and less fertile than an A. Clay, aluminum, iron, and some nutrients have been washed or leached from the E horizon. The E is apparent in the Miami soil (Plate 2) and the Hosmer soil (Plate 12). In Crosby soil (Plate 3), part of the E remains undisturbed below the plowed (Ap) horizon, especially on the right side of the profile.

B Horizon

This is the mineral horizon below an A and/or E horizon and is commonly called the subsoil.

It also has distinctive characteristics caused by:

- Accumulation of clay, iron, aluminum, or a combination of these;
- Prismatic or blocky structure;
- Reddish or brownish colours;
- Weak cementation, resulting in brittle material, or
- Some combination of the factors.

Together, A, E, and B horizons are called the solum. If a soil lacks B and E horizons, as in recent alluvial materials, the A horizon alone is the solum.

C Horizon

The material that cannot be designated as A, E, or B horizon because it lacks soil development is the C horizon, often called the substratum. The C horizon may be the weathered rock material immediately below the solum, or it may be material that was moved by ice, water, or wind. In most soils, the C horizon is like the material from which the overlying horizons formed. If this is the case, we can say that the C is the parent material of the A, E, and B horizons. Sometimes the C horizon material is different from the material in which the A, E, and B horizons formed.

R Horizon

This horizon is the consolidated (hard) bedrock, such as limestone, sandstone, or shale. The depth to bedrock in Indiana varies from a few inches to several hundred feet.

RECOGNIZING SOIL DEVELOPMENT

In soil evaluation, contestants must learn to recognise soil d evelopment because very weak soil development is one of the criteria for identifying alluvium and local overwash. Soil development is recognised by profile differences in texture, structure, colour, or, some combination of these properties.

As pointed out in the section on soil formation processes, clay is formed during soil development and tends to move down in the profile. Thus in a developed soil, there is usually more clay in the B horizon (subsoil) than there is in the A or E horizon or in the C horizon. Also, this B horizon usually has blocky or prismatic soil structure development. Soil particles are grouped into peds that can be recognised in a chipped (not cut) portion of the profile; and they can be removed by gently probing into the soil with a knife or spatula. These ped surfaces are often covered with clay films or clay skins. To learn to recognise soil structure, you probably will have to work with an experienced person such as a soil evaluation coach or a soil scientist.

In well and moderately well drained soils, the B horizons are usually more red or brown than the A, E, or C horizons. This is caused by accumulation of iron released from soil minerals to form red or brown iron oxide minerals. In somewhat poorly and poorly drained soils, almost all the profile has either gray colours due to reduction and removal of iron, or mottled colours due to migration of iron over short distances within a horizon to cause depletion in the gray areas and concentration in the brownish or reddish spots. Soil development is reflected by profile differences in texture, colour, structure, or cementation, or by changes in several of these properties. In some sandy soils, profile differences are in colour only. Therefore, careful observation and experience are needed to judge soil development. It is often helpful to carry some material from one site to another during practice sessions so that materials can be compared directly.

SOIL ORGANIC MATTER ON SOIL PROPERTIES

Organic matter affects both the chemical and physical properties of the soil and its overall health. Properties influenced by organic matter include: soil structure; moisture holding capacity; diversity and activity of soil organisms, both those that are beneficial and harmful to crop production; and nutrient availability. It also influen ces the effects of chemical amendments, fertilizers, pesticides and herbicides. This chapter focuses on those properties related to soil moisture and water quality.

INEFFICIENT USE OF RAINWATER

Drylands may have low crop yields not only because rainfall is irregular or insufficient, but also because significant proportions of rainfall, up to 40 per cent, may disappear as run-off. This poor utilization of rainfall is partly the result of natural phenomena (relief, slope, rainfall intensity), but also of inadequate land management practices (*i.e.* burning of crop residues, excessive tillage, eliminating hedges, etc.) that reduce organic matter levels, destroy soil structure, eliminate beneficial soil fauna and do not favour water infiltration. However, water "lost" as run-off for one farmer is not lost for other water users downstream as it is used for recharging groundwater and river flows.

Where rainfall lands on the soil surface, a fraction infiltrates into the soil to replenish the soil water or flows through to recharge the groundwater. Another fraction may run off as overland flow and the remaining fraction evaporates back into the atmosphere directly from unprotected soil surfaces and from plant leaves.

The above-mentioned processes do not occur at the same moment, but some are instantaneous (run-off), taking place during a rainfall event, while others are continuous (evaporation and transpiration).

To minimize the impact of drought, soil needs to capture the rainwater that falls on it, store as much of that water as possible for future plant use, and allow for plant roots to penetrate and proliferate. Problems with or constraints on one or several of these conditions cause soil moisture to be one of the main limiting factors for crop growth. The capacity of soil to retain and release water depends on a broad range of factors such as soil texture, soil depth, soil architecture (physical structure including pores), organic matter content and biological activity. However, appropriate soil management can improve this capacity.

Practices that increase soil moisture content can be categorized in three groups:

1. Those that increase water infiltration;
2. Those that manage soil evaporation; and,
3. Those that increase soil moisture storage capacities. All three are related to soil organic matter.

In order to create a drought-resistant soil, it is necessary to understand the most important factors influencing soil moisture.

INCREASED SOIL MOISTURE

Organic matter influences the physical conditions of a soil in several ways. Plant residues that cover the soil surface protect the soil from sealing and crusting by raindrop impact, thereby enhancing rainwater infiltration and reducing run-off.

Surface infiltration depends on a number of factors including aggregation and stability, pore continuity and stability, the existence of cracks, and the soil surface condition. Increased organic matter contributes indirectly to soil porosity (via increased soil faunal activity). Fresh organic matter stimulates the activity of macrofauna such as earthworms, which create burrows lined with the glue-like secretion from their bodies and are intermittently filled with worm cast material.

The proportion of rainwater that infiltrates into the soil depends on the amount of soil cover provided. The figure shows that on bare soils (cover = 0 tonnes/ha) run-off and thus soil erosion is greater than when the soil is protected with mulch. Crop residues left on the soil surface lead to improved soil aggregation and porosity, and an increase in the number of macropores, and thus to greater infiltration rates.

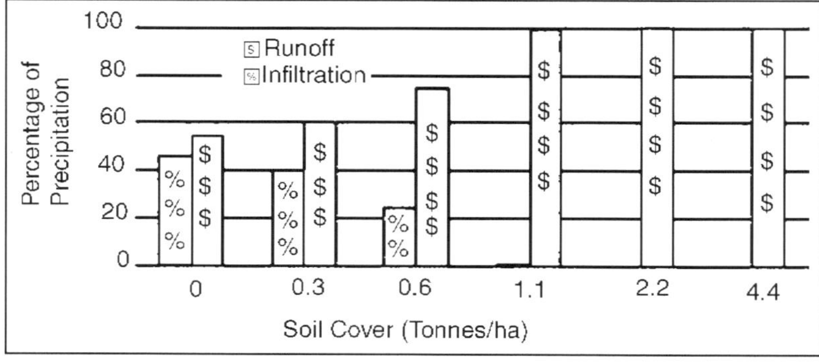

Fig. Effect of Amount of Soil Cover on Rainwater Run-off and Infiltration

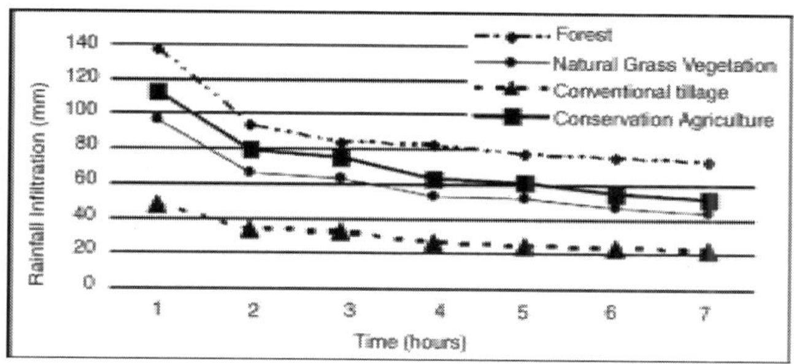

Fig. Water Infiltration Under Different Types of Management

Increased levels of organic matter and associated soil fauna lead to greater pore space with the immediate result that water infiltrates more readily and can be held in the soil. The improved pore space is a consequence of the bioturbating activities of earthworms and other macro-organisms and channels left in the soil by decayed plant roots.

On a site in southern Brazil, rainwater infiltration increased from 20 mm/h under conventional tillage to 45 mm/h under no tillage. Over a long period, improved organic matter promoted good soil structure and macroporosity. Water infiltrates easily, similar to forest soils (Figure).

The consequence of increased water infiltration combined with a higher organic matter content is increased soil storage of water. Organic matter contributes to the stability of soil aggregates and pores through the bonding or adhesion properties of organic materials, such as bacterial waste products, organic gels, fungal hyphae and worm secretions and casts. Moreover, organic matter intimately mixed with mineral soil materials has a considerable influence in increasing moisture holding capacity. Especially in the topsoil, where the organic matter content is greater, more water can be stored.

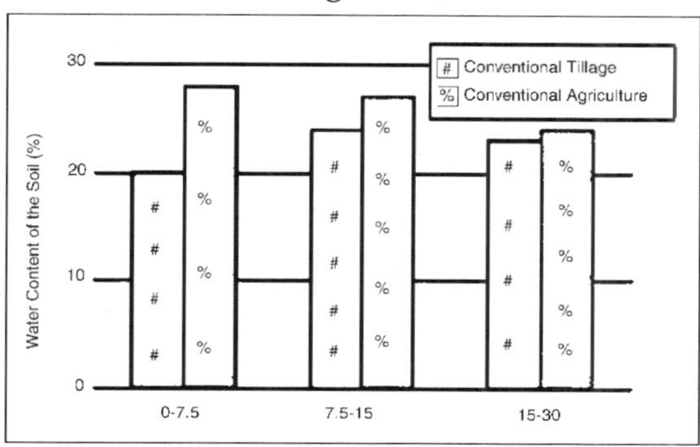

Fig. Quantity of Water Stored in the Soil Under Conventional Tillage and Conservation Agriculture

The quality of the crop residues, in particular their chemical composition, determines the effect on soil structure and aggregation. Blair *et al.* report a rapid breakdown of medic (Medicago truncatula) and rice

(Oryza sativa) straw residues resulting in a rapid increase in soil aggregate stability through the release of many soilbinding components. As these compounds undergo further breakd own, they will be lost from the system resulting in a decline in soil aggregate stability over time. The slow release of soil-binding agents from flemingia (Flemingia macrophylla) residues resulted in a slower but more sustained increase in the stability of soil aggregates. This indicates that continual release of soil-binding compounds from plant residues is necessary for continual increases in soil aggregate stability to occur.

Elliot and Lynch showed that soil aggregation is caused primarily by polysaccharide production in situations where residues have a low N content. There is a strong relationship between soil carbon content and aggregate size. An increase in soil carbon content led to a 134-per cent increase in aggregates of more than 2 mm and a 38-per cent decrease in aggregates of less than 0.25 mm. The active fraction of soil C is the primary factor controlling aggregate breakdown.

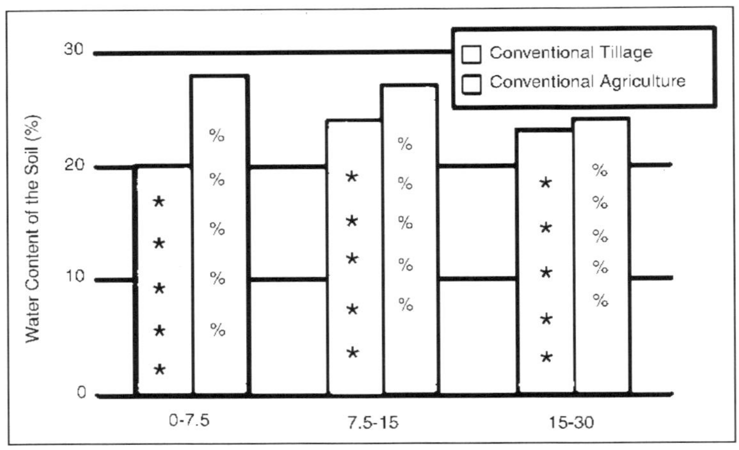

Fig. Effect of Different Soil Covers on in-Soil Storage of Water

In addition, although they do not live long and new ones replace them annually, the hyphae of actinomycetes and fungi play an important role in connecting soil particles. Gupta and Germida showed a reduction in soil macroaggregates correlated strongly with a decline in fungal hyphae after six years of continuous cultivation.

The in-soil storage of water depends not only on the type of land preparation but also on the type of cover or previous vegetation on the soil. Figure indicates the effect of burning vegetation on the amount of water stored in the soil.

Conserving fallow vegetation as a cover on the soil surface, and thus reducing evaporation, results in 4 per cent more water in the soil. This is roughly equivalent to 8 mm of additional rainfall. This amount of extra water can make the difference between wilting and survival of a crop during temporary dry periods. A study conducted in 1999 in Guatemala, Honduras and Nicaragua to evaluate the resilience of agro-ecosystems showed that 3-15 per cent more water was stored in the soil under more ecologically sound practices. Unger showed that high wheat-residue levels resulted in increased storage of fallow precipitation, which subsequently produced higher sorghum grain yields. High residue levels of 8-12 tonnes/ha resulted in about 80-90 mm more stored soil water at planting and about 2.0 tonnes/ha more of sorghum grain yield compared to no residue management.

Table. Average Soil Depth at Which Moisture Starts, and Difference in Moisture Stored

Country	Agro-ecologically sound practices cm	Conventional practices cm	Difference (%)
Honduras	9.98	10.28	2.9
Guatemala	2.44	2.99	15.0
Nicaragua	15.81	17.80	11.2

The addition of organic matter to the soil usually increases the water holding capacity of the soil. This is because the addition of organic matter increases the number of micropores and macropores in the soil either by "gluing" soil particles together or by creating favourable living conditions for soil organisms. Certain types of soil organic matter can hold up to 20 times their weight in water. Hudson showed that for each 1-per cent increase in soil organic matter, the available water holding capacity in the soil increased by 3.7 per cent.

Table. Economy of Irrigation Water Through Soil Cover, the Brazilian Cerrados

Country	Agro ecologically sound practices (cm)	Conventional practices	Difference (%)
Honduras	9.98	10.28	2.9
Guatemala	2.44	2.99	15.0
Nicaragua	15.81	17.80	11.2

Soil water is held by adhesive and cohesive forces within the soil and an increase in pore space will lead to an increase in water holding capacity of the soil. As a consequence, less irrigation water is needed to irrigate the same crop.

NUTRIENTS IN THE SOIL ORGANIC MATTER

Soil organic matter not only stores nutrients in the soil, but is also a direct source of nutrients. Some of the world's most fertile soils tend to contain high amounts of organic matter. Soil organic matter includes all organic (or carbon-containing) substances within the soil.

Soil organic matter includes:

- Living organisms (soil biomass)
- The remains of microorganisms that once inhabited the soil
- The remains of plants and animals
- Organic compounds that have been decomposed within the soil and, over thousands of years, reduced to complex and relatively stable substances commonly called humus.

As organic matter decomposes in the soil, it may be lost through several avenues. Since organic matter performs many functions in the soil, it is important to maintain soil organic matter by adding fresh sources of animal and plant residues, especially in the tropics where the decomposition of organic residues is continuous throughout the year.

Important Functions of Organic Matter

Although surface soils usually contain only 1-6 per cent organic matter, soil organic matter performs very important functions in the soil. Soil organic matter:

- Acts as a binding agent for mineral particles.
  - This is responsible for producing friable (easily crumbled) surface soils.
- Increases the amount of water that a soil may hold.
- Provides food for organisms that inhabit the soil.
- Humus is an integral component of organic matter because it is fairly stable and resistant to further decomposition.
  - Humus is brown or black and gives soils its dark colour.
  - Like clay particles, humus is an important source of plant nutrients.

SOIL WATER

In nutrient management, a proper balance between soil water and soil air is critical since both water and air are required by most processes that release nutrients into the soil. Soil water is particularly important in nutrient management. In addition to sustaining all life on Earth, soil water provides a pool of dissolved nutrients that are readily available for plant uptake. Therefore, it is important to maintain proper levels of soil moisture.

Soil water is important for three special reasons:

- The presence of water is essential for the all life on Earth, including the lives of plants and organisms in the soil.
- Water is a necessary for the weathering of soil. Areas with high rainfall typically have highly weathered soils. Since soils vary in

their degree of weathering, it is expected that soils have been affected by different amounts of water.

- Soil water is the medium from which all plant nutrients are assimilated by plants. Soil water, sometimes referred to as the soil solution, contains dissolved organic and inorganic substances and transports dissolved nutrients, such as nitrogen, phosphorus, potassium, and calcium, to the plant roots for absorption.

The amount of water in the soil is dependent upon two factors:

- First, soil water is intimately related to the climate, or the long term precipitation patterns, of an area.
- Secondly, the amount of water in the soil depends upon how much water a soil may hold.

Soil water holding capacity

Before we discuss the capacity of soils to hold water, we must understand the concept of capillarity.

Capillarity

- Water molecules behave in two ways:
  - *Cohesion Force*: Because of cohesion forces, water molecules are attracted to one another. Cohesion causes water molecules to stick to one another and form water droplets.
  - *Adhesion Force*: This force is responsible for the attraction between water and solid surfaces. For example, a drop of water can stick to a glass surface as the result of adhesion.
- Water also exhibits a property of surface tension:
  - Water surfaces behave in an unusual way because of cohesion. Since water molecules are more attracted to other water molecules as opposed to air particles, water surfaces behave like expandable films. This phenomenon is what makes it possible for certain insects to walk along water surfaces.
- Capillary Action:
  - Capillary action, also referred to as capillary motion or capillarity, is a combination of cohesion/adhesion and surface tension forces.
  - Capillary action is demonstrated by the upward movement of water through a narrow tube against the force of gravity.
  - Capillary action occurs when the adhesive intermolecular forces between a liquid, such as water, and the solid surface of the tube are stronger than the cohesive intermolecular forces between water molecules.
  - As the result of capillarity, a concave meniscus (or curved, U-shaped surface) forms where the liquid is in contact with a vertical surface.
  - Capillary rise is the height to which the water rises within the

tube, and decreases as the width of the tube increases. Thus, the narrower the tube, the water will rise to a greater height.

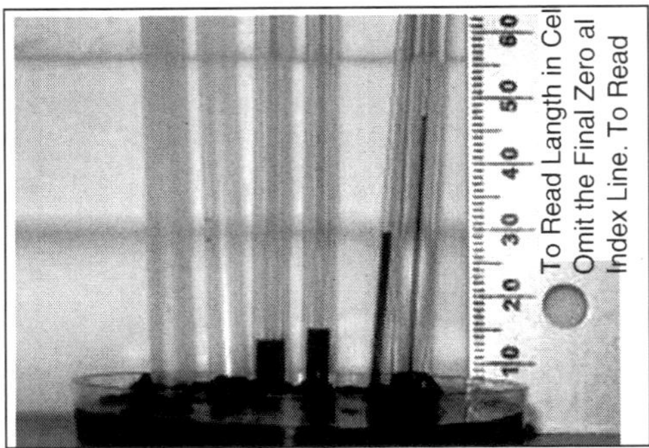

Fig. Capillary Rise in Tubes of Varied Widths.

This picture demonstrates the phenomenon of capillary rise. As you can see, the liquid rises to the greatest height in the narrowest tube (at far right), whereas capillary rise is lowest in the widest tube (at far left). Although easily demonstrated by simple experiments using tubes, capillary action occurs in soils. Smaller pores that exist in finely-textured soils have a greater capacity to hold and retain water than coarser soils with larger pores.

Capillary action is the same effect that causes porous materials, such as sponges, to soak up liquids.

- Capillarity is the primary force that enables the soil to retain water, as well as to regulate its movement.
  - The phenomenon of capillarity also occurs in the soil. In the same way that water moves upwards through a tube against the force of gravity; water moves upwards through soil pores, or the spaces between soil particles.
  - The height to which the water rises is dependent upon pore size. As a result, the smaller the soil pores, the higher the capillary rise.
  - Finely-textured soils, typically have smaller pores than coarsely-textured soils. Therefore, finely-textured soils have a greater ability to hold and retain water in the soil in the inter-particle spaces. We refer to the pores between small clay particles as micropores. In contrast, the larger pore spacing between lager particles, such as sand, are called macropores.
  - In addition to water retention, capillarity in soil also enables the upward and horizontal movement of water within the soil profile, as opposed to downward movement caused by gravity. This upward and horizontal movement occurs when lower soil layers have more moisture than the upper soil layers and is important because it may be absorbed by roots.

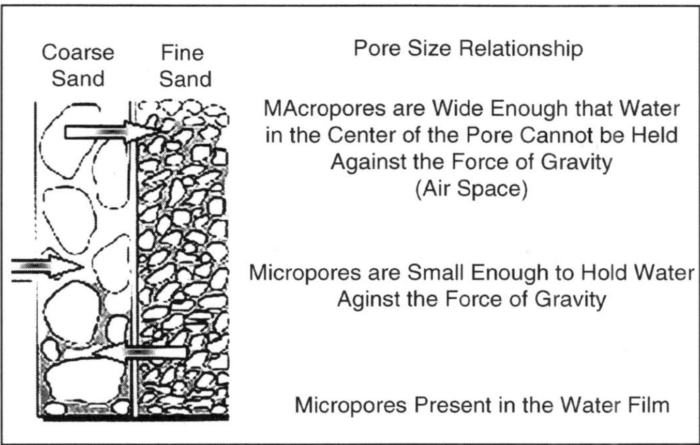

Fig. This Picture Shows how more Water may be Held between Finer Particles Against the Force of Gravity, as Compared to Coarser Particles. As a Result, Finer-textured Soils have Greater Water Holding Capacities.

Water Holding Capacity

Since water is held within the pores of the soil, the water holding capacity depends on capillary action and the size of the pores that exist between soil particles. Sandy soils have large particles and large pores. However, large pores do not have a great ability to hold water. As a result, sandy soils drain excessively. On the other hand, clayey soils have small particles and small pores. Since small pores have a greater ability to hold water, clayey soils tend to have high water holding capacity.

SOIL AIR

In nutrient management, soil aeration influences the availability of many nutrients. Particularly, soil air is needed by many of the microorganisms that release plant nutrients to the soil. An appropriate balance between soil air and soil water must be maintained since soil air is displaced by soil water. Air can fill soil pores as water drains or is removed from a soil pore by evaporation or root absorption. The network of pores within the soil aerates, or ventilates, the soil. This aeration network becomes blocked when water enters soil pores. Not only are both soil air and soil water very dynamic parts of soil, but both are often inversely related:

- An increase in soil water content often causes a reduction in soil aeration.
- Likewise, reducing soil water content may mean an increase in soil aeration.
- Since plant roots require water and oxygen (from the air in pore spaces), maintaining the balance between root and aeration and soil water availability is a critical aspect of managing crop plants.

Soil air is very different than the above-ground atmosphere. A significant difference is between the levels of carbon dioxide. Since the soil contains high amounts of carbon dioxide, oxygen levels may become limited. Since plants must have oxygen to live, it is important to allow proper aeration in the soil. See Tables 3 and 4 for references to soil air composition.

Table. Comparison between Soil air and Atmospheric Composition

Component	Soil Air (%)	Atmosphere (%)
N_2	79.2	79.0
O_2	20.6	20.9
CO_2	0.25	0.03

Soil Atmosphere

- The soil atmosphere is not uniform throughout the soil because there can be localised pockets of air.
- The relative humidity of soil air is close to 100 per cent, unlike most atmospheric humidity.
- Air in the soil often contains several hundred times more carbon dioxide.

3

Mineral Function of Soils

The mineral fraction of soils is derived largely from weathering of the underlying parent material, which may consist of consolidated bedrock or unconsolidated superficial deposits. Consolidated bedrock can be classified into one of three categories—igneous, sedimentary or metamorphic.

Igneous rocks are derived from the consolidation of molten magma, either within the Earth's crust or at its surface, whereas sedimentary rocks are the product of cycles of weathering, erosion and deposition operating at the surface, and metamorphic rocks originate from the alteration of any other rock type by high temperature and/or pressure but without melting. Unconsolidated superficial deposits are just as variable as solid bedrock materials and are often classified according to the nature of their depositional environment, for example riverine alluvium, glacial till and lacustrine clays. These deposits are often indurated or cemented to some extent by calcareous, siliceous or iron-rich components. In addition to the parent material origin of mineral material, it can also be added to soils by movement from upslope, by aeolian transport or by atmospheric fall-out of materials such as volcanic ash.

Soil minerals occur in a wide variety of forms, but these can be arranged into groups on the basis of their chemical composition and the structural arrangement of their constituent elements. By far the most abundant group in soils and their parent materials is the *silicate* group. The fundamental building block of all silicate minerals is the silicon-oxygen (Si-O) tetrahedron, which consists of a central silicon ion (Si^{4+}) surrounded by four closely spaced oxygen ions (O^{2-}).

Because the silicon ion has four units of positive charge and each oxygen ion has two units of negative charge, each discrete tetrahedron possesses four units of negative charge.

This allows the tetrahedra to link together in a variety of characteristic structural arrangements which forms the basis of their classification. The potential range of silicate minerals is further widened by substitution of one ion for another within the mineral structure, a process known as *isomorphous substitution*. One of the most common substitutions involves replacement of some Si^{4+}, in Si-O tetrahedra, by aluminium ions (Al^{3+}). Other examples include the substitution of Al^{3+} by magnesium ions (Mg^{2+}) or iron (Fe^{3+} or Fe^{2+}). When ions with the same charge or valency are exchanged, the mineral structure remains electrically neutral. If, however, ions with different valencies are exchanged, there will be a charge imbalance; frequently this results in an excess negative charge, as in the examples above.

In this situation, electrical neutrality is achieved either by incorporation of additional cations such as calcium, magnesium, potassium or sodium (Ca^{2+}, Mg^{2+}, K^+, Na^+) into the crystal lattice, or by structural rearrangements that allow internal compensation of charge. The most common silicate minerals and their associated structures are discussed briefly below.

Framework silicates or tectosilicates consist of a three-dimensional lattice of Si-O tetrahedra linked through their corners. The two main mineral groups within this category are quartz and feldspars, which are common minerals in many soils. Quartz consists simply of Si-O tetrahedra linked through the O^{2-} ions, and consequently there are twice as many O^{2-} ions as Si^{4+} ions in the mineral structure, which has the general formula $(SiO_2)n$. Unlike quartz, feldspars contain significant amounts of Al^{3+} ions, originating from isomorphous substitution of Si^{4+} ions, and base cations, especially Ca^{2+}, Na^+ and K^+, which satisfy the residual negative charges.

There are two main groups of feldspar minerals—potassium feldspars such as orthoclase and microcline, and plagioclase feldspars which form a continuous series of minerals between the sodium end member, albite, and the calcium end member, anorthite.

Chain silicates or *inosilicates* comprise Si-O tetrahedra linked together to form a continuous chain structure of which there are two types. First there is the single chain where the Si-O tetrahedra are linked by sharing two out of the three basal O^{2-} ions, and second there is the double chain where tetrahedra are linked by sharing all three basal O^{2-} ions to form a hexagonal arrangement.

Common minerals in these two categories are the pyroxene family and the amphibole family respectively. In chain silicates the chains themselves are linked together by a variety of cations including Ca^{2+}, Mg^{2+}, Fe^{2+}, Na^+ and Al^{3+}. Minerals in this group are very reactive and more easily weathered than the framework silicates. *Orthosilicates* and *ring-silicates* display a wide range of structures but, in comparison with other groups, their occurrence in soils is relatively minor.

Some minerals in this group are very reactive and therefore particularly susceptible to weathering, whereas others are very resistant. This category can be divided into two groups—*neso-silicates* and *soro-silicates*. In the first group the Si-O tetrahedra occur as separate units, with no shared O^{2-} ions, and are linked by metallic cations. Examples include olivine and garnet. In soro-silicates the tetrahedra form separate groups in which they share one or more of their O^{2-} ions; when the group is formed by sharing O^{2-} ions by more than two Si-O tetrahedra, a ring structure results. Examples of soro-silicates are beryl and cordierite. *Sheet silicates* or *phyllosilicates* are perhaps the most important group of minerals in soils, playing an important role in the development of many physical and chemical soil characteristics. Unlike the other mineral groups, many sheet silicate minerals occur in the smallest grain-size categories; consequently they are often known as the *clay minerals*. Minerals in this group are made up of various combinations of three fundamental sheet structures:

- Si-O (siloxane) sheet in which the Si-O tetrahedra are linked in a hexagonal arrangement,
- Al-OH (gibbsite) sheet in which the Al^{3+} ions are surrounded by six closely packed hydroxyl (OH^-) ions to form an octahedral structure,
- Mg-OH (brucite) sheet whose structure is similar to that of the gibbsite sheet but contains Mg^{2+} ions rather than Al^{3+}.

Classification of sheet silicate minerals is based on three criteria—the ratio of the above sheets within a unit layer of the mineral structure, the interlayer or basal spacing between unit layers, and the interlayer components or species. The sheet silicate minerals most commonly found in soils include the micas, illite, chlorite, kaolinite, vermiculite and the smectites.

The most frequently occurring micaceous minerals are muscovite and biotite. Both of these are 2:1 layer silicates with an interlayer spacing of approximately 10 Å (1.0 nm) and K^+ as the dominant interlayer component. Illite is very similar in composition and structure to micaceous minerals and is sometimes known as hydrous mica.

Chlorite has a 2:1:1 layer structure (two siloxane sheets to one brucite sheet, forming a biotite mica layer, to one further brucite sheet which occupies the interlayer position) with an interlayer spacing of about 14 Å (1.4 nm). Kaolinite has a 1:1 layer structure and an interlayer spacing of around 7 Å (0.7 nm). Layer units of kaolinite are attracted by hydrogen bonding and the crystals have a characteristic hexagonal appearance. Vermiculite is a 2:1 layer silicate with a basal spacing of about 14 Å (1.4nm) and interlayer positions occupied by Mg^{2+} ions and water molecules. The most common smectitic mineral in soils is montmorillonite. This is a 2:1 layer silicate with interlayer positions dominated by Na^+ and Ca^{2+} ions, and water molecules. Although its basal spacing is approximately 14 Å (1.4 nm), this varies somewhat due to its shrink-swell characteristics which occur on wetting and drying.

Other sheet silicate minerals sometimes found in soils include halloysite, imogolite and allophane, and mixed layer or interstratified minerals. Halloysite is similar in many ways to kaolinite, but its crystal layers are curved to form a tubular structure. Imogolite and allophane are gel-like hydrated alumino-silicate minerals with a poorly crystalline tube- or thread-like structure. Mixed layer minerals result from the interlayering of more than one sheet silicate mineral. Interlayering may be regular, where there is a regular repetition of the different mineral layers, or random where there is no clear pattern of repetition. These minerals may exist as intergrades which represent stages of transition between one mineral and another during weathering. Examples of mixed layer minerals are mica- or illite-vermiculite (hydrobiotite), illite-montmorillonite and kaolinite-montmorillonite.

In addition to the silicate minerals are the *non-silicate minerals*. These are often of only minor occurrence in soils and are usually referred to as *accessory minerals,* although they play a significant role in the development of some soil characteristics. Among the most common non-silicate minerals in soils are the free oxides and hydroxides of iron, aluminium and manganese; these may exist as crystalline or amorphous forms.

Iron minerals include goethite, ferrihydrite, hematite, lepidocrocite, maghemite and magnetite. Aluminium minerals include gibbsite and boehmite, while examples of manganese minerals are birnessite and pyrolusite. Other non-silicate minerals include calcite ($CaCO_3$), which occurs in soils developed from calcareous parent materials, and anatase (TiO_2) and amorphous silica, which are often found in soils developed on recent volcanic deposits.

Organic Components

Soil organic matter is derived from a number of sources, of which the most important is usually plant litter. This consists of a variety of plant debris including leaves, stems, flowers, twigs, bark, and the larger branches and trunks of trees. Other organic components include plant roots, root exudates, soil organisms together with their faecal remains and metabolites, and organic substances washed into the soil from vegetation.

Soil organisms are responsible for the physical comminution and bio-chemical decomposition, and incorporation of organic matter. They exhibit tremendous diversity in terms of their numbers, size and morphology, and also vary dramatically in their function, mode of nutrition and environmental tolerance. Organisms can be referred to as *producers, consumers* or *decomposers.*

Producers, such as plants, fix carbon from atmospheric carbon dioxide (CO_2) during photosynthesis, consumers feed on plants and other organisms, and decomposers utilise carbon from organic material, returning it to the atmosphere as CO_2 and other gaseous by-products and mineralising nutrients to their original ionic form. Soil organisms can also be classified according to their size *micro-organisms,* which include both microflora and microfauna, *mesofauna* and *macrofauna*. The classification of soil organisms according to their mode of nutrition is based on their sources of carbon and energy. Essentially, they obtain carbon from either organic material or inorganic sources (largely CO_2); these two groups are known as *heterotrophs* and *autotrophs* respectively. They can also obtain energy either from light (photoheterotrophs or photoautotrophs) or from chemical oxidation (chemoheterotrophs or chemoautotrophs). In addition to their carbon and energy sources, soil organisms, especially bacteria, vary in their oxygen requirements. *Aerobes* have a direct requirement for oxygen, *facultative anaerobes* normally require oxygen but may adapt to oxygen deficit by utilising nitrate and other inorganic compounds as electron receptors, and *obligate anaerobes* require an absence of oxygen as it is toxic to them. Before looking at the composition of organic material, the various groups of micro-organisms, mesofauna and macrofauna will be briefly considered.

Four main groups of micro-organisms can be recognised—bacteria and actinomycetes, fungi, algae and protozoa. Bacteria are very small organisms (1-5 ìm) and are often round or rod-like in shape. They tend to exist in thin films of water surrounding soil particles, often reproducing with tremendous rapidity, with numbers in the range of 10^6-10^9g having been widely reported. Many species secrete polysaccharide gums, which facilitate aggregation. Bacteria demonstrate great biochemical versatility in their ability to decompose a wide range of materials under a variety of conditions. For example, *Pseudomonas* sp. can metabolise a range of chemicals including pesticides, *Nitrobacter* sp. derives its energy from the oxidation of nitrite to nitrate, *Thiobacillus ferrooxidans* acquires energy from oxidation of reduced sulphur compounds and Fe^{2+}, and *Rhizobium* sp. forms nitrogen-fixing nodules on the roots of leguminous plants.

Actinomycetes consist of fine branching filaments (about 1 ìm in diameter) during their vegetative stage, when they are similar in appearance to fungi. During reproduction, however, the filaments undergo fragmentation, sometimes forming dense colonies. For this reason actinomycetes are often classified as highly evolved and complex bacteria. Fungi produce filamentous structures (hyphae) which are about 0.5-1.0 ìm in diameter and which grow into a dense network or mycelium. They are generally less numerous than bacteria in soils (1-4×10^5 g), although they are common in acidic soils where they can be responsible for 60-80 per cent of organic matter decomposition. All fungi are heterotrophic, living mostly in the surface layers where they play an important role in the development of soil structure. Mycorrhizal fungi live symbiotically in plant tissue, removing carbon in return for nutrients such as phosphorus.

Algae are photosynthetic organisms and are therefore confined largely to the soil surface. They include Cyanophaceae (blue-green algae) and Chlorophaceae (green algae), the former playing a major role in development of the nitrogen status of soils. Protozoa are microfauna, unlike the previous groups which are microflora. They are uni- or non-cellular organisms of 5-40 ìm in length and live in thin water films surrounding soil particles. They can be divided into amoeboid forms with silica or chitin sheaths, and rotifers with better differentiated cell structure. Protozoa are very numerous (often $> 10^4$ g) and are particularly important in controlling the numbers of bacteria and fungi, on which they feed.

As in the case of micro-organisms, four main groups of mesofauna can be recognised—nematodes, arthropods, annelids and molluscs. Nematodes are unsegmented worms, and next to protozoa are the smallest of the soil fauna, being 0.5-1.0 mm in length. They are plentiful in soil and litter, feeding on plant remains, roots, bacteria and sometimes on protozoa.

Arthropods can be divided into a number of categories, the most significant numerically being acari (mites) and collembola (springtails). Both of these feed on plant litter, bacteria and fungi, with acari being particularly common in acidic litter where they may constitute 80 per cent of soil fauna.

Other arthropod groups include myriapods, which are dominated by chilopods (centipedes) which are carnivorous, and diplopods (millipedes) which are herbivorous. Arthropods also include isopods (woodlice), beetles, insect larvae, ants and termites. Termites are prevalent in tropical soils and are particularly active soil mixers, producing structures at the surface (termitaria) up to several metres in height.

Arthropod populations can be particularly high, for example 220,000 m^2 in old grassland soils. Annelids consist largely of enchytraeid worms (potworms) and lumbriscid worms (earthworms). Enchytraeids are small (0.1-5.0 cm in length) and have a thread-like appearance. They feed on algae, fungi, bacteria and soil organic matter, and can occur in populations as high as 200,000 m^2. Lumbriscid worms are probably more important than any other soil invertebrate in the decomposition and mixing of organic matter, and their numbers can exceed 800 m^2, although they do not tolerate highly acidic conditions, and during prolonged drought or heat they burrow deeply into the soil and become inactive.

Some species, such as *Lumbricus rubellus,* live only in the surface litter layer, but most migrate between organic and mineral horizons, ingesting both constituents and ensuring thorough mixing of the soil. Large populations under grassland can consume 90 ha, and the casts which are produced are important in the development of soil structure. The main mesofaunal mollusc groups are slugs and snails (gastropods) which feed on plants, fungi and faecal remains, but these are often limited in biomass to 20-45 g/m^2. Macrofauna include larger molluscs, beetles and larger insect larvae, together with larger vertebrates such as moles, rabbits, foxes and badgers. These often burrow deeply into the soil, feeding on smaller organisms in their path. Moles in particular are voracious feeders and consume large numbers of earthworms, insect larvae and slugs.

In terms of its composition, organic material consists predominantly of carbon, hydrogen, oxygen and nitrogen, with carbon providing the framework for organic structures; most organic residues in soils contain around 45-55 per cent carbon by weight. The elements which constitute organic material are arranged to form a variety of compounds, some of which have very complex structures. The basic building block of many organic compounds, however, is the carbon tetrahedron, where a central carbon atom with a valency of four links with other elements, notably hydrogen, oxygen and nitrogen; this structural arrangement is very similar to the basic building block of silicate minerals.

In some instances, numerous tetrahedra link together to form extremely large molecules known as *polymers*. Linkages can produce both linear and cyclic molecules, and the number of potential combinations and structural arrangements is almost endless. Consequently, organic structures and their variety are considerably more complex than those of mineral material and are less well understood, although the majority of organic compounds can, however, be isolated. The main groups of compounds are carbohydrates, proteins and amino acids, and lignin, along with smaller amounts of fats, waxes, pigments and resins. The proportions of these compounds vary according to the type of organic matter and its stage of decomposition.

Essentially, carbohydrates are hydrates of carbon with a general formula $Cx\ (H_2O)y$, although some contain other elements such as nitrogen and sulphur. Carbohydrates are the main constituent of plant material at 60-90 per cent of dry mass, and in soils 5-30 per cent of the carbon exists in this form. The simplest and most easily broken down carbohydrate compounds in soils are the sugars and starches. These comprise relatively small molecules such as glucose (a simple sugar), which is a monosaccharide. More complex, and thus more resistant to breakdown, are hemi-cellulose (a pentose sugar polymer) and cellulose (a β (1-4) glucose polymer).

These are polysaccharides which consist of monosaccharide units joined by C-O-C links. Cellulose is the most abundant carbohydrate in plants, where it may constitute more than 40 per cent of the carbon. It forms the resistant fibres of plants, the breakdown of which is catalysed by enzymes (cellulases) secreted by micro-organisms.

Breakdown ultimately converts the polysaccharide compounds into monosaccharide compounds which can then be assimilated directly by the micro-organisms. Chitin is a polysaccharide of similar composition to cellulose, but it also possesses some nitrogen in the form of amino groups. It is a common constituent of insect cuticles and is also often present in fungi.

Proteins and amino acids are composed of about 50-55 per cent carbon, 20-25 per cent oxygen, 15-20 per cent nitrogen and 6.5-7.5 per cent hydrogen, although some contain small amounts of phosphorus and sulphur. They are an important source of nitrogen, with around 20-50 per cent of all organic nitrogen in soils existing as amino acids. Although they are rapidly metabolised in soils, they can persist for long time periods due to adsorption onto other constituents, such as clay, or by combination with more resistant organic components such as lignin or tannin.

Lignin is a particularly resistant component of soil organic material, and constitutes 15-35 per cent of the supporting tissues of plants. Essentially it is a complex phenolic polymer which displays a high degree of aromaticity in the development of cyclic, benzene ring structures. The aromatisation and large number of C-C bonds are the main cause of its resistance to decomposition. As a result of organic matter decomposition, fresh organic components are gradually converted, via a range of fermented products, to a material known as *humus,* which represents the end-product of this process. Humus is finely divided and amorphous with no cellular structure. Essentially it consists of insoluble, heterogeneous polymers and has an approximate composition, on an ash-free basis, of 44-53 per cent carbon, 40-47 per cent oxygen, 3.5-5.5 per cent hydrogen and 1.5-3.5 per cent nitrogen.

Water

Soil water is derived from two principal sources—precipitation and ground water. Precipitation arrives at the soil surface in various forms; although rain, snow and hail are the main types, fog and mist may provide significant amounts of moisture, particularly in coastal and upland areas. The proportion of precipitation that reaches the ground surface depends largely on the nature and density of vegetation cover.

On surfaces devoid of vegetation, precipitation reaches the soil directly, but on vegetated surfaces a significant proportion of precipitation can be intercepted; much of this ultimately reaches the soil as canopy throughfall and stemflow, while some is returned to the atmosphere by evapouration. On reaching the surface, water can either infiltrate the soil or, if the rate of arrival of water at the surface exceeds the rate of infiltration, it will run off over the surface. The precipitation that infiltrates the soil will either be lost via evapotranspiration or drainage, or will be retained by forces which hold it in place or allow it to move only very slowly. Ground water can be derived by lateral movement from upslope, or by upward movement from the underlying rock strata.

The composition of soil water is a particularly dynamic characteristic, varying dramatically even over short time periods. This behaviour arises from the intimate association between the water, small mineral and organic particles (clay and humus) and plant roots, which can involve the exchange of ions between these components. Soil water contains a number of dissolved solid and gaseous constituents, many of which exist in mobile ionic form, and a variety of suspended solid components. Base cations (Ca^{2+}, Mg^{2+}, K^+, Na^+, NH^{4+}) may be derived from a number of sources. They are present in the atmosphere and are later dissolved in precipitation; in maritime areas, for example, precipitation often contains significant quantities of marine salts which are particularly rich in sodium and chloride ions (Na^+ and Cl^-). As well as being deposited in the soil directly, ions accumulate on the surfaces of vegetation and are subsequently washed off into the soil. Cations can also be derived from mineral weathering and organic matter decomposition, entering the soil directly or via the soil exchange system; these processes play an important role in the buffering of soil acidity. In agricultural systems, lime and fertilizers provide yet another source of base cations, particularly Ca^{2+}, K^+ and NH^{4+}.

The concentration of hydrogen ions (H^+) in soil water is a measure of its acidity, which is expressed in terms of *pH*. A major source of such acidity is carbon dioxide (CO_2) which is derived from the atmosphere, where it is dissolved in precipitation, and from the soil air where it is a product of soil organism respiration. Carbon dioxide dissolved in water Unpolluted rain water in equilibrium with atmospheric CO^2 has a pH of about 5.6, whereas soil water in equilibrium with CO_2 in soil air is usually more acidic, with pH values often below 5.0; this is because CO_2 levels in soil air are considerably greater than in the atmosphere.

Another source of acidity in soil water derives from industrial and urban emissions. In addition to these pollutants, organic acids derived from decaying organic material are an important source of soil acidity. H^+ is also released by plants in exchange for nutrient base cations, and as part of the process of nitrification where NH^{4+} is converted to NO^{3-}.

Iron and aluminium are also important constituents of soil water, particularly under acidic conditions. They are derived from mineral weathering and may exist in the soil solution either as ions (Fe^{2+} and Al^{3+}) or in the form of soluble organo-metallic complexes; significant quantities of dissolved organic carbon may also exist in this form. In some circumstances aluminium may be mobilised by mineral acids, such as sulphuric and nitric acid deposited in acid precipitation. As in the case of cations, the anions in soil water are derived from a number of sources. Nitrate and phosphate ions (NO_3 and $PO_4{}^{3-}$) are produced by mineralisation processes during organic matter decomposition, and are also derived from fertilizers. Chloride ions (Cl^-) and, to a lesser extent, sulphate ions ($SO_4{}^{2-}$) originate from atmospheric sources, including airborne marine salts and acid deposition. Bicarbonate ions (HCO_3) originate largely from the dissociation of HCO_3 as mentioned above, and are associated with mineral weathering, especially in soils developed from carbonate-rich parent materials. In addition to the major dissolved constituents of soil water, there are other dissolved components although these are usually relatively minor and local in their occurrence. These include organic material and silica, together with a number of pollutants such as heavy metals and radionuclides.

Soil water contains not only dissolved solids but also a number of suspended constituents. These include small particles of mineral and organic material, which often result in discolouration and increased turbidity of soil water. Similarly, precipitates may accumulate in soil water, usually as a result of chemical changes as the water migrates through the soil. For example, orange-brown precipitates of insoluble ferric iron compounds can sometimes be observed where water, in which iron compounds usually exist in the soluble ferrous form, becomes more oxygenated. The chemical characteristics of water can also influence the behaviour of fine sediments, particularly clays, in suspension.

In waters with low solute concentrations, or significant quantities of monovalent cations such as Na^+, the clays tend to remain in a dispersed state and can be held in suspension for long periods of time. In waters with high solute concentrations, however, particularly where concentrations of divalent cations such as Ca^{2+} are high, the clays may undergo rapid flocculation and settling.

Air

Air and water have a reciprocal arrangement in terms of their occupancy of soil pore space; in saturated soils, air content is low, whereas in dry soils the pore spaces are largely air-filled. Changes in water and air content are particularly dynamic because much of the water present in a saturated soil drains away rapidly, while heavy rainfall can quickly bring the soil back to saturation.

The gaseous constituents of soil air are derived largely from the atmosphere, the respiration and metabolism of soil organisms, and from the evapouration of soil moisture. Soil air is continuous with the atmosphere provided that the soil surface is not sealed due to compaction or crusting, and such continuity ensures the free movement and exchange of gases.

The gases move along gradients of partial pressure, so that oxygen will tend to migrate from the atmosphere where its partial pressure is high into the soil where it is low. Conversely, carbon dioxide and water vapour will tend to migrate from the soil into the atmosphere. Movement of gases may occur by diffusion, mass flow or in dissolved form, with diffusion being by far the most significant of these processes.

The atmosphere contains approximately 78 per cent nitrogen, 21 per cent oxygen and 0.03 per cent carbon dioxide by volume. In comparison, soil air contains similar amounts of nitrogen, slightly less oxygen and more carbon dioxide.

The balance between levels of oxygen and carbon dioxide in soil air depends largely on the rate of respiration of soil organisms and on the diffusivity characteristics of the soil. Respiration rates often vary seasonally in response to the availability of organic substances for consumption, and to variations in temperature and moisture conditions. Diffusion of gases occurs most effectively in dry soils with large and interconnected pores.

The presence of water tends to reduce pore continuity and consequently the soil becomes increasingly oxygen deficient or *anaerobic*. In addition to carbon dioxide, organisms release other gases into the soil, including methane (CH_4) and hydrogen (H_2), as a result of organic matter decomposition.

SOIL MINERALS AND CHANGE IN MINERAL COMPOSITION

Weathering is the principal process that acts upon the earth's primary minerals to form the smaller and finer particles that we call "soil." In terms of nutrient management, the process of weathering greatly influences the availability of plant nutrients.

Initially, as soil particles begin to weather, primary minerals release nutrients into the soil. As these particles decrease in size, the soil is also able to retain greater amounts of nutrients. Ultimately, however, the capacity to hold and retain nutrients is greatly reduced in highly weathered soils, since most nutrients have been lost due to leaching.

There are two types of weathering: physical weathering and chemical weathering. Differences in weathering patterns are the reason why there is a great range in soil particle size. Boulders are much less weathered than gravel. In return, gravel is much less weathered than clay particles. Clay particles may even weather into other materials, such as iron and aluminum oxides, which are generally resistant to further weathering. In the tropics, chemical weathering is very important. Since the climate is typically warm and moist year-round, it provides a suitable environment for continuous chemical weathering to occur. Over time, with sufficient amounts of rainfall and warm temperatures, mineral particles weather into smaller and smaller soil particles. As a result, tropical soils tend to be highly weathered soils. Table provides a list of common primary, secondary minerals, aluminum and iron oxides, and amorphous materials.

PHYSICAL WEATHERING

Physical weathering is a process that breaks up and disintegrates parent rock, or primary minerals, within the earth. In the tropics, physical weathering is caused by the wetting and drying of rocks; erosion; actions of plants and animals; or the falling, smashing, or breaking of rock materials into smaller pieces.

Chemical Weathering

Chemical weathering is important in nutrient management since the resulting soil particles retain and supply nutrients. However, when highly weathered, the soil loses much of its nutrients due to excessive leaching. Thus, highly weathered soils tend to be infertile soils, while moderately weathered soils are generally more fertile.

Once parent rock has broken down into smaller pieces, another process acts upon the rock. This process is chemical weathering. Chemical weathering involves the change, or transformation, of primary minerals into secondary minerals. Secondary minerals serve as the basic building blocks of the small particles with the soil. As a result, new materials may be synthesised, residual material may accumulate from materials (such as oxides) which cannot be furthered weathered, or materials can be lost as the result of leaching.

Table. Important Primary Minerals and Weathered Materials

Important Minerals and Weathered Materials of Basalt Rock	
Primary Minerals of Basalt Rock	• Plagioclase Feldspar • Olivine • Augite Others: magnetite, apatite, ilmenite
Secondary Minerals	• Smectite, such as montmorillonite (less weathered) • Kaolin, such as halloysite (more weathered)
Iron Oxides	• Hematite • Goethite • Magnetite • Maghemite • Lepidocrosite • Ferrihydride
Aluminum Oxide	• Gibbsite
Amorphous Minerals	• Allophane • Imogolite
Amorphous Substances	No specific names given

MINOR NUTRIENTS LEVELS IN SOILS

Magnesium levels in soils are usually adequate, though prolonged use of lime can deplete magnesium ions in exchangeable cation sites through competition by calcium ions. Occasional use of dolomite, $CaMg(CO_3)_2$, or magnesium sulfate, supplies magnesium.

Calcium levels are also usually adequate, though prolonged use of ammonium sulfate can deplete levels. The use of lime or superphosphate usually supplies adequate calcium.

Iron and manganese are usually abundant in soils, but become very insoluble at higher pH levels. Plants probably absorb iron as a soluble "chelate" formed with a chelating (complex forming) agent excreted by the roots. Manganese levels are in a delicate balance between deficiency and toxicity, and can cause problems in some acid soils.

Other minor nutrients are *usually* present in adequate amounts, but can be deficient in particular regions. Local deficiencies can sometimes be identified by the symptoms shown by plants (*e.g.* yellowing of leaves caused by a deficiency of the magnesium needed to make chlorophyll), or by chemical analysis of the soil or plant material. Once the problem is identified then fertilizers containing the trace element can be applied.

If a deficiency problem is suspected it may be simpler to use a fertilizer containing a mixture of the commonly lacking trace nutrients. Fertilizers are commonly labelled with four numbers, which give the percentages of N, P, K and S; *e.g.* superphosphate is labelled 0 9 0 12 (9 per cent P and 12 per cent S). A variety of mixed fertilizers with different compositions are available; one available with all the major nutrients has composition 7 5 7 14. One important factor in choosing fertilizers is how rapidly the nutrients become available to plants. Soluble inorganic compounds, such as urea, ammonium nitrate and potassium phosphate, are immediately available, and are commonly used for hydroponic solutions, and as foliar feeding solutions (plants can absorb soluble nutrients from solutions sprayed on their leaves). It is often better if the nutrients are released into the soil slowly throughout the growing season of the plant, and *slow release* fertilizers are used for this reason.

Sometimes nutrient deficiencies affect animals which eat the plants, rather than the plants themselves. Many of the soils of the central North Island plateau contain little cobalt, which is not required by plants, but is required by animals to make vitamin B_{12}.

Cobalt is added as a fertilizer supplement and is taken up by the plants thereby providing the animals' requirements. There is also some evidence that added selenium, which is also present in low concentration in some New Zealand soils, may improve animal health though it is not needed by plants.

NUTRIENT STATUS OF SOILS ANALYSIS

The nutrient status of soils can be determined by chemical analysis, in which the amount of a particular element in a known weight of soil is measured. However, not all of the element present in the soil may be available to plants, so forms of analysis which attempt to mimic the ability of plants to extract nutrients from the soil are commonly used.

For example, the soil may be extracted with a citric acid solution, and the amount of the nutrient in the solution determined. The best extracting solutions have been found by trials in which the results of analysis using different extracting solutions are compared with plant growth observed in field trials.

The available cationic nutrients are generally determined by extracting a known weight of air-dried soil with an extracting solution, and then determining the amounts of K^+, Mg^{2+}, Ca^{2+}, etc, present using atomic absorption or atomic emission spectrophotometers, which measure the amount of light characteristic of a particular element absorbed or emitted when the solution is sprayed into a flame.

The other important parameter that needs to be monitored is the pH of the soil. Soil becomes progressively more acidic over time as a soil is leached by rainwater saturated with carbon dioxide.

The cations are slowly replaced by hydrogen ions in reverse order to the strength of their binding by the soil, and carried away to the oceans by the percolating water. The singly charged ions are lost first, then the doubly and finally the triply charged ions, and the soil progressively becomes more acid in character. Over geological time, even the aluminium and silica is leached away, and the surface of the land is steadily lowered. This exposes a fresh layer of subsoil to the weathering process, which releases new supplies of nutrients into the newly forming soil in a process of continuous renewal of the topsoil with its nutrients. The physical movement of soil particles, by gravity, wind or water in the process of *erosion* can greatly speed this process, especially on slopes. For gardeners and farmers the soil acidity is a useful indicator of availability of nutrients. This property is often referred to as *soil pH,* though more strictly it refers to the pH of the soil solution. The "soil pH" is readily measured, and gives useful information about the state of a soil. In *acid soils* (pH < 6), many of the exchangeable sites in the clay minerals and humus are occupied by hydrogen or aluminium ions.

Remaining nutrient cations are readily released by the soil, but there may not be much of these nutrients present. In very acidic soils, cations such as Fe^{3+} and Al^{3+} appear in the soil solution, and these are quite toxic to many plants. At the other end of the pH range, in basic soils (pH > 8), which are generally derived from limestone, the cation nutrients generally become less available. Plant species have evolved to grow in soils with a particular pH range, and may not thrive in soils with acidities outside this range. For examples; rhododendrums grow best in strongly acid soils, most trees, shrubs and grasses in slightly acid soils, and many common vegetable crops grow best in neutral to slightly basic soils. The acidity of a soil can be adjusted to suit the type of plants being grown. Most New Zealand native forest plants have developed to grow under the prevailing acid soil conditions, but the soil pH needs to be raised for optimum growth of most crop plants.

The use of acid fertilizers, such as superphosphate, makes the soil more acidic. Soil is normally made less acidic by using lime (finely ground limestone, $CaCO_3$). For faster reaction, "slaked lime" can be used. This is made by heating limestone to form calcium oxide (quicklime, CaO), and then reacting this with water to form calcium hydroxide, $Ca(OH)_2$.

Lime also adds the necessary nutrient calcium, and improves the texture of acid soils (makes them less "sticky") by replacing the H of the -OH groups between the layers of the clay minerals by Ca^{2+} ions. If the soil is deficient in magnesium, the mineral *dolomite,* $CaMg(CO_3)_2$ can be used instead of lime to raise the pH. n the rare occasions when the soil pH needs to be lowered, *e.g.* to grow rhododendrons, this can be achieved by applying sulfur or ammonium sulphate.

These are oxidised, releasing hydrogen ions and so making the soil more acidic:

$$S(s) + 3O_2(g) + 2H_2O(l) \rightarrow 2SO_4{}^{2-}(aq) + 4H^+(aq)\ NH_4{}^+(aq)$$
$$+ O_2(g) + H_2O(l) \rightarrow NO_3{}^-(aq) + 6H^+(aq)$$

GENERAL ANALYSIS

First look at where the soil is, and where it came from. Was the soil formed in position by weathering of the local substrate? What is the nature of the local substrate (sedimentary or igneous rock, volcanic ash, sand, - -)? Or has it been deposited from elsewhere; by the sea, by a river, by the wind, or by human activity? Or was it formed in peat, which is the largely organic residue of an ancient swamp? Much can be learned about a soil just by looking at it and touching it. Look at the soil profile of a freshly cut bank, or dig a hole about 40 cm deep, and study the soil profile. Note the colour and feel of the soil, and the way these change with depth.

Note the amount of organic material (plant debris) present—high for peaty soils to low for clay subsoils. Soils with much organic material are generally dark coloured, soils with little organic material but high in iron(III) oxide are yellow and those low in iron are grey. Rub a sample between your fingers; is it fine and smooth or coarse and gritty? Does it stick to your fingers, or rub off cleanly? Shake a sample of the soil up with water and let it settle; sand and coarse unweathered mineral particles will settle rapidly, fine clay particles will stay suspended as "muddy" water for a long time, while coarse organic material may even float. Try to find what can be deduced from these properties for your local soil.

Animal Life

Carefully crumble about 100 g of freshly collected soil and look for larger animals, such as earthworms and grass grubs. Spread the soil over a sieve (or piece of mesh curtain). Place the sieve over a piece of paper, cover, and leave in a dark place overnight. Try to identify the livestock which falls onto the paper.

Soil pH

Place about 10 g (a teaspoonful) of soil in about 100 ml of distilled water (or fresh rainwater) in a clean stoppered bottle and shake occasionally for about 5 minutes. Allow the solids to settle then gently press a glass electrode into the sediment layer and use a pH meter to determine the pH of the solution. If you do not have access to a pH meter, dip a non-bleeding pH "stick" into the solution and compare the colour developed with the chart; or filter the solution through fine filter paper (or centrifuge the solution) to produce a colourless solution then add a few drops of universal indicator, and compare the colour with the chart to determine the pH. What types of plants will grow well in soil with this "soil pH"?

Available Phosphate Determination

Phosphate is usually analysed by using a reaction with molybdic acid in the presence of a reducing agent. An intensely blue coloured compound (phosphomolybdic acid) is formed, and the colour intensity is proportional to the amount of phosphate present. The colour is compared with that developed in solutions containing a known amount of phosphate. For analysis of an inhomogeneous material such as soil the collection of a *representative sample* is extremely important. The phosphate level will vary from place to place, and with depth. You could choose a sample from some marked place (and the results would apply only to that place) or collect samples from typical places in an area, *e.g.* from hill-sides, from the level, near trees, etc, etc, and then thoroughly mix them, to get an average value for the area. The usefulness of the result will depend on how representative of the whole area your sample is. Collect a soil sample (or samples). Allow to air dry for a few days so that reproducible results can be obtained. Roughly grind the soil using a pestle and mortar (thoroughly mix if average samples are being used). Accurately weigh out about 0.5 g of soil and place in a stoppered flask, Add 100 mL of the extractant solution A, and shake (or stir with a magnetic stirrer) for 30 minutes. Filter through a fine filter paper.

Pipette 50 mL of the filtrate into a 100 mL volumetric flask, and add about 30 mL of water. Add 8 mL of reagent C and make up to 100 mL. The blue colour is developed over 10 minutes and is stable for 24 hours. Pipette 0, 5, 10, 15, 20 and 25 mL of the standard solution E into labelled 100 mL volumetric flasks and fill to about 80 mL with water. Add 8 mL of reagent C and make up to 100 mL. Compare the intensity of colour of the unknown and standard solutions.

If you have a colourimeter, use a blue filter. If you have a spectrophotometer, the maximum absorption is at 880 nm, though a secondary maximum at 660 nm can be used. If you have no instrument, look sideways through the flasks at white paper and try to interpolate the unknown among the standards (or better, pour each solution into a *marked* test tube, filling them all to the same depth, and look down the tubes at white paper. Estimate the extractable phosphate content of the solution and hence of the soil: ìg P g^{-1} of soil = 4 x ìg P ml^{-1} in solution.

Phosphate Retention by Soil

The ability of soils to bind phosphate gives useful information about the likely effectiveness of addition of phosphate fertilizers. In this test the soil is shaken with a standard phosphate solution, and the amount of phosphate adsorbed is determined by analysing the residual solution. The test is most useful for comparing different soils. Accurately weigh about 5 g of ground air dried soil and place in a stoppered container. Add 25 mL of reagent F (which contains 1 mg P mL^{-1}) and shake or stir for 24 hours. Filter through fine filter paper. Pipette 5 mL of the filtrate into a 500 mL volumetric flask, and make up to the mark. This solution would contain 10 ìg P mL^{-1} if the soil had not adsorbed any phosphate. Prepare calibration solutions by pipetting 10 and 25 mL of solution F into 50 mL volumetric flasks, and make up to the mark. Pipette 5 mL of each of these solutions, and of solution F, into labelled 500 mL volumetric flasks and make up to the mark. Pipette 5 ml of the analysis solution, and of each of the calibration solutions, into separate labelled 100 ml volumetric flasks, add about 80 ml of water and 8 ml of reagent C to each and make up to the mark. Compare the blue colours after 30 min as for the previous test. The calibration solutions correspond to 80, 50 and 0 per cent phosphate retention, respectively.

Solutions for Phosphate Analysis

- Reagent A (Truog's reagent). Dissolve 7.5 g of $(NH_4)_2SO_4$ plus 5 mL of 1 mol L^{-1} H_2SO_4 in 2.5 L of distilled water.
- Reagent B. Dissolve 30 g of ammonium molybdate in 1 L of water (warm to dissolve, but do not heat above 60° C). Dissolve 0.667 g of antimony potassium tartrate in 250 mL water. Add these solutions to 1.25 L of 5 mol l^{-1} H_2SO_4 (353 mL conc. H_2SO_4 made up to 1.25 L). Mix and make up to 2.5 L and store in a dark bottle. If antimony potassium tartrate is not available, the test can still be applied, but the colour development is slower. Allow the solutions to stand for at least 30 minutes, or heat in a boiling water bath for 10 minutes before comparing the colours.
- Reagent C (Murphy and Riley reagent). Take as much of reagent B as required and dissolve 1 g of ascorbic acid (vitamin C) for every 100 mL. Mix thoroughly. This solution does not keep for more than 24 hours.
- Reagent D; Phosphate standard stock solution (100 μg P mL^{-1}). Dissolve 0.220 g of potassium dihydrogen phosphate (KH_2PO_4) in water and make up to 500 mL in a volumetric flask. Add 0.5 mL of toluene if the solution is to be kept for any length of time.
- Reagent E; Phosphate working solution (1 μg P mL^{-1}). Pipette out 5 mL of the stock solution D and make up to 500 mL. Prepare as required.
- Reagent F (1 mg P mL^{-1} at pH 4.6). Dissolve 4.4 g of KH_2PO_4 and 16.4 g of anhydrous sodium acetate in water in a 1 L volumetric flask, add 11.5 mL of glacial acetic acid, and make up to the mark.

ESSENTIAL MINERAL NUTRIENTS OF PLANT GROWTH ON SOILS

Solution culture experiments have established a number of facts basic to plant growth on soils—that plant roots must have a supply of oxygen, that solid soil particles and micro-organisms, while sometimes benecial, are not essential, that in addition to C, H and O some 14 or 15 chemical elements are essential for plant growth, and that all of these essential elements may be supplied to plant roots as simple ions of inorganic salts in solution and must be supplied in adequate but non-toxic amounts. A chemical element is regarded as essential if, in its absence, a plant cannot complete its life cycle. An alternative criterion has also been suggested—that the element is part of an essential plant constituent or molecule. So far, only the rst criterion has been used to establish essentiality although the latter has proved useful as a guide to further research, as, for example, in the association of Mn with laccase activity and Ni with urease activity.

'Essential mineral nutrients' or simply 'essential nutrients' include all those chemical elements which are normally absorbed from the soil solution by higher plants. They exclude C, H, and O which comprise more than 99 per cent of metabolically active leaves and some 90–95 per cent of their dry matter, but, paradoxically, they include N which does not occur in any soil mineral and is supplied to some plants from air. Essential mineral nutrients comprise less than 1 per cent of fresh mass in active leaves and 5–10 per cent of their dry matter; yet within this small proportion, all 14 or 15 elements must be present in adequate amounts for healthy growth and effective reproduction.

FUNCTIONS OF NUTRIENTS

Table. A Summary of notional values for concentrations of nutrient elements (on a Dry Mass Basis) considered as just adequate for maximum palnt growth, and accompanying data on the relative number of atoms of those elements witnin a plant (referenced to Mo=l). A broad distribution between plant requirements for micronutrients (trace elements) and micronutrients is than evident. Mobility form leaves is meant to reflect overall impressions chemical analaysis of nutrient budgets for whole plants and organs.

Element **Mg kg^{-1}**	**Concentration** **% (Dry**	**Number of of Atoms mass)**	**Relative**	**Relative Number** **Atoms**
		Micronutrients		
Molyhdenum	0.1	—	1	Variable
Nivkel	0.1	—	2	Mobile
Copper	3	—	50	Variable
Zinc	20	—	300	Variable
Mangancse	20	—	400	Immobilc
Boron	12	—	1,000	Immobilc
Iron	100	—	2,000	Immobilc
Chlorine	100	—	3,000	Mobile
Sodium	500	—	20,000	Mobile
		Macronuutrients		
Sulphur	—	0.1	30,000	Variable
Calcium	—	0.2	50,000	Immobilc
Phosphorus	—	0.2	60,000	Mobile
Magnesium	—	0.2	80,000	Variable
Potassium	—	1.0	250,000	Mobile
Nitrogen	—	2.5	1,300,000	Mobile

Historically, essential nutrients have been classied in one of two groups based on amounts required by plants, namely macronutrients or micronutrients, and that convention has been adopted for convenience in Table. This distinction reflects historical sequence and experimental difûculties in discovering essential nutrients. Following advances in nine-teenth century chemistry, essentiality of macronutrients was relatively easy to prove, but essentiality of micronutrients (except Fe) was elusive, requiring great care in eliminating contamination from macronutrient salts, from water and from other environmental sources. The classication remains in common use, and was therefore used in Table but the distinction between macro and micro is somewhat arbitrary. In terms of chemistry and function, an alternative grouping is outlined below.

Group 1: N, S

N and S are covalently bound in organic compounds in reduced states. They are essential components of proteins providing reactive groups for interaction with metabolites and other nutrients. They are generally absorbed as oxyanions which, except for small amounts of sulphate in a few organic com-pounds, must be reduced before use.

Group 2: P, B

P and B are covalently bound in organic compounds in their fully oxidised states. P is present in phospholipids of cell membranes, nucleic acids of chromosomes and a large number of intermediary metabolites including many in which it plays a crucial role in bioenergetics. P is always present as mono- or polyorthophosphate which behaves as a weak acid. B, like P, is thought to function only in its fully oxidised form; B oxyacid is much weaker and is largely undissociated at the pH of cells. B complexes strongly with hydroxyls in adjacent *cis*-diol conguration in sugars and their derivatives such as those in the hemicellulose of cell walls, but key metabolic or structural functions still remain unknown.

Group 3: K, Ca, Mg, Na, Cl

K, Ca, Mg, Na and Cl are present primarily in ionic form as either free ions in solution or as entities reversibly adsorbed by electrostatic forces on charged sites. K, Ca and Mg are required in relatively large amounts while Na and Cl, though often present in large amounts, are required only in trace amounts. Group 3 nutrients function in osmotic adjustment of cell activities and in enzyme activation, probably by modifying the shape and orientation of substrates and enzymes. Ca and to a lesser extent Mg also stabilise cell structures such as membranes and cell walls, probably via their divalent positive charge which forms cross-links with negative charges on cell structures. Some Ca binds strongly and reversibly to the protein calmodulin which in turn activates several enzymes and controls Ca transport. Some Mg is structurally bound in chlorophyll (a strict 1:1 stoichiometry of Mg atoms to chlorophyll molecules is universal)

Group 4: Mn, Fe, Co, Ni, Cu, Zn, Mo

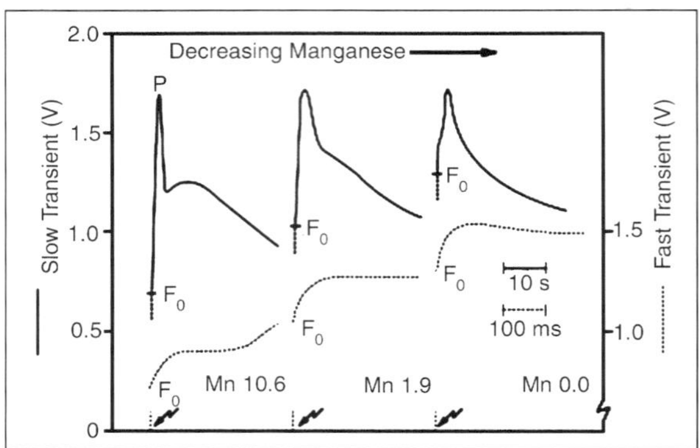

Fig. Induction Kinetics for *in Vivo* Chlorophyll *a* Fluorescence Change Dramatically According to Leaf Mn Status of Wheat Leaves.

Constnat yield fluorescence (F_0) increases but variable fluorescence (F_v; where $F_v = P–F_0$) decreases with decrease in leaf Mn. A ratio of F_0/F_v increases abruptly as leaf Mn drops to a critical level, and provides an early indication of impending Mn deficiency. Preplanting soil dressings corresponding to transients on leaf samples were, left to right, 10.6, 1.9 and 0.0 kg $MnSO_4$ ha^{-1}. Fe, Co, Ni, Cu, Zn, Mo and, to some extent, Mn are tightly bound in proteins as metalloproteins or, in the case of Co, which is only required in plants xing N_2, in a coenzyme. All are transition series elements, and with the exception of Zn all show variation in valency state with attendant variation in physiological effect. Group 4 elements govern a wide range of reactions including many oxidation–reductions in which all except Zn can act as electron carriers by undergoing reversible oxidation–reduction.

All are micronutrients, and all except Mn form strong complexes with organic and amino acids. Mn is exceptional in that it resembles Group 3 nutrients in being present largely as a free divalent ion which can replace Mg in the activation of a number of enzymes. But some Mn is tightly bound within the water-splitting apparatus of photo-system II and in superoxide dismutase, an enzyme system that helps dissipate harmful effects of singlet oxygen formation. Mn deciency results in highly charac-teristic induction kinetics for *in vivo* chlorophyll *a* fluorescence and can be used as a diagnostic tool in combination with chemical assay.

4

Soil Biology and Nutrition

The aim of management should be to create balanced organic matter and mineral budgets. It should ensure that, over several years (a complete crop rotation), soil organic matter is not depleted and that nutrients added equal or exceed those removed by cropping or lost in various ways.

Various approaches to good management. These complement the aspects and the advice on saline water management given by Kandiah and that on water harvesting offered by Critchley and Siegert.

When managing organic matter farmers should recognise that the effects of animal and human manure, sewage sludge and plant residues last longer than those of green manure crops. Green manures, though valuable, usually last only one or two seasons because they are incorporated before they are mature and lignified. The longer-term beneficial effects of manures on soil organisms.

Juo and Payne advise: 'In spite of the many proven benefits of soil organic matter, its management and recycling in an intensified, modem agro-ecosystem must necessarily revolve around two fundamental characteristics of the system, namely, the availability of organic material at the farm level, and the economic incentive for conserving and recycling organic matter'. To this may be added consideration of the benefits and costs of treating and transporting human wastes from cities. On-site organic matter, organic material brought in from outside and industrially-produced fertilizers will be used according to their benefit cost ratios and the attitudes of governments and crop managers.

Tillage reduces the frequency of VAM, at least in invertebrates. Tillage reduces earthworm populations 10-30-fold, both killing them directly and destroying their burrows. Haines and Uren show that differences in earthworm populations between tillage treatments and stubble burning are reflected by 50 per cent differences in numbers of soil pores.

Tillage also effects the location, numbers and activity of micro-organisms within the soil, depending on the type of tillage. For example a mouldboard plough that inverts topsoil has more effect than minimal tillage. Doran found microbial biomass and potentially-mineralisable nitrogen to be 54 per cent and 37 per cent higher respectively in the top 7.5 cm of no-till than ploughed soil. These would be critical differences if translated to the semi-arid tropics. Microbial and fungal biomasses may be 20-70 per cent higher after 1 and 2 years no-till. However, conventional tillage, by burying topsoil, may cause microbial populations and their activities to be higher than under no-till at depths of say 7.5 to 15 cm. Differences in soil organisms (*e.g.*, microbial biomass) under differing tillage treatments are seasonal, generally being greatest in the period when the weather is most favourable for soil organism activity and plant growth. Rasmussen and Collins summarise many data on the impact of tillage on soil properties and conclude that stubble-mulching with zero tillage conserves up to 2 per cent more organic matter per year than ploughing. Cropping patterns and rotations affect soil nutrition. Diversity of cropping increases the number and variety of soil organisms and reduces pests and diseases.

Soil acidification and salination are extreme cases of nutrient imbalance and, unlike other deficiencies, cannot be corrected simply by adding mineral fertilizers. The management techniques to reduce acidification are:

- To reduce the net proton production of the system by minimising nitrate leaching (a special problem in seasonally wet-and-dry climates);
- To avoid using ammonium fertilizers and reduce the accumulation of organic matter;
- To lime the soil.

Liming is a high cost option while adoption of perennial species (to avoid seasonal accumulation of organic matter and nitrate leaching) is less costly. By contrast, salinity is usually managed by strategic leaching. This is done with the sparing use of low-salt water when both soil moisture.

INORGANIC NUTRITION IN SOIL

Inorganic nutrients occur in soil as ions and minerals, *e.g.*, as oxides, silicates and phosphates, both adsorbed onto the surface of clay particles and organic matter, and in solution.

A large proportion of some nutrients, most notably nitrogen, is found in organic matter in all but the most highly degraded soils, so organic matter and the organisms associated with it are given.

Clay particles, because of their crystalline structures, carry an inherent electrical charge. This results in attractive forces (mainly van der Waals forces) and repulsive forces (electrostatic forces) which give clay species their particular characteristics. The inherent surface charge also causes a layer of associated ions to align next to the solid particles forming a so-called diffuse double layer because it consists of a relatively inexchangeable layer (the Stem layer) closest to the surface of the particle and an outer, readily exchangeable layer, of varying thickness, called the diffuse layer. Sposito explains this more fully. In the present context, it is appropriate to describe briefly four aspects of the physico-electrical structure of soils. These are: the inherent exchange capacity; the linkage between soil particles and soil organic matter, and its effects on soil stability; ions in solution and their solubility; and the ion imbalances (from the perspective of crop growth) associated with acidity and salinity.

Adsorption is the net accumulation of matter at the interface between a solid phase and an aqueous solution. Readily exchangeable ions are those so loosely adsorbed to clay or other particles that they are replaced easily by leaching with an electrolyte solution. The ion exchange capacity of soil is the number of moles of adsorbed ion charge which can be displaced from a unit mass of soil; in most applications, this refers to readily exchangeable ions. Sposito notes that 'Much controversy exists over the surface chemical significance of ion exchange capacities'. The maximum surface charge measurement indicates a soil's potential to adsorb ions while its actual capacity, which is more relevant agriculturally, has a lesser value. The representative cation exchange capacities (CECs) for selected soil orders. It is notable that though the CEC of each order ranges widely, the predominant soils in wet-and-dry climates have low CECs.

The readily exchangeable proportion of the CEC varies with soil pH: the proportion available to readily-exchangeable bases decreases from about 1 at pH 8 to 0.5 at pH 6 to perhaps 0.2 at pH 4.5. Below pH 6 an increasing proportion is taken up by various aluminium ions and complexes, which are variably toxic to plants. The electrical conductivity and CEC of a soil are related to its clay content. Similarly, because of the electrical charge of the clay particles, a high but variable percentage of the soil organic matter is bound to them. It may be as high as 90 per cent, the two postulated main ways that organic matter is bonded to clay. These are weak anion exchange and strongly-held ligand exchange which is a form of chemical bonding.

It follows from the electrically-charged nature of clay and organic matter and the high likelihood of their association that both contribute to the nutrient-holding capacity and stability of soils.

Plants take up most minerals as inorganic nutrients from the soil solution (except legumes, which may fix dinitrogen gas directly from the soil air). An element is available (to the plant) if it is present as, or can be transformed into, a free ion, and it is within the plant root zone. Most elements move within physical proximity of roots through soil water movement and into the plant through evapotranspiration.

Diffusion along concentration gradients is important for less-mobile ions such as phosphorus, particularly where soil solution concentrations are weak and root densities are high (*i.e.*, the transport path is short). Sposito and others give calculated values for the diffusivity of nutrients. Diffusion times range from 1 day for an ion to move 3 mm (which is comparable with the time it would take to move by convection in the mass flow of water) for nitrate to about 200 days for potassium, magnesium and molybdenum, and to thousands of days for other nutrients. The concentrations of nutrients in solution fluctuate daily and seasonally. The most dynamic are nitrate and ammonium ions, which are interconverted by bacteria. Fluctuations can be explained by the effects of temperature and soil water on mineralisation (breaking down organic matter to release ammonium), immobilisation (the reverse) and nitrification (conversion of organic matter to nitrate, which is stable and highly soluble), and by the effects of rainfall leaching nitrate to depth.

In seasonally wet-and-dry cropping systems, there is a flush of nitrate following the start of the wet season and nitrate-N may accumulate in the topsoil by capillary rise of water during the dry season. Within the major seasonal patterns driven by soil water, Rochester *et al.*, found that the soil nitrate cycle lagged three months behind changes in temperature.

Low soil pH affects plant roots directly because of the effects of the hydrogen ion concentration on root membrane integrity and exchange capacity. Acidity also affects roots indirectly in two ways. It alters the availability of ions in the soil solution (possibly making available toxic aluminium species, relatively toxic to plants).

It also affects mineralisation through the protons competing with cations for dissolved ligands and surface charged groups. Soil pH also affects micro-organisms, and thus the speed of transformations, for example, those between nitrate and ammonium.

Poor plant growth on acid soils may thus be caused directly by hydrogen ions, by toxicities of aluminium or manganese, or through deficiencies of calcium, magnesium, potassium, phosphorus, nitrogen or trace elements. Chase *et al* describe these relationships for sandy Sahelian soils. Variation in pH across distances of 15 m can be as much as pH 4.5 to 7.5, with associated decreases in aluminium and hydrogen ions, and increases in crop productivity. The critical pH at which crop growth is affected varies with crop, cultivar and soil type. Critical levels may be as high as pH 5 to 5.5 for less tolerant plants in soils with soluble sources of aluminium but otherwise can be as low as pH 3.9 to 4.

Saline soils and saline-sodic soils have electrical conductivities greater than 4 dS/m. The saturated soil extract of saline soils contains more than 15 per cent exchangeable sodium; that of sodic soil less than 15 per cent.

The pH values of both kinds of soils range between 7 and 10.5 directly reflecting the amount of sodium bicarbonate present, this being completely dissolved in moist soil. The sensitivities of crop yields to salt. Longer lists of plant tolerance to salt are given by Maas and Rhoades.

SOIL NUTRIENTS

The strong canola yield response to S fertilizer on deficient soils has been well established in western Canada. Under extreme S deficiency, canola response to S fertilizer is dramatic.

However, canola response to S fertilizer varies greatly, depending on:

- Soil sulphate levels (amount, spatial and temporal distribution)
- Availability of other nutrients (especially N, P and possibly boron)
- Soil moisture
- Amount, type, and method of S fertilizer applied

Since canola absorbs S from soil as sulphate, the soil sulphate content affects the yield response to applied S fertilizer. Researchers in the 1960-70s established a relationship between soil sulphate level and canola yield. Water soluble soil sulphate was found to be a good measure of available S for canola growth. Canola generally responded to S fertilizer when the sulphate content to 60 cm (2') was less than 22 to 34 kg/ha (20 to 30 lb/ac). However, soil testing to determine sulphate content and the likelihood of yield response to applied S fertilizer has not been consistently successful.

For example, two field experiments in Manitoba in 1964 failed to show a rapeseed yield response to sulphate fertilizer on deficient soil while two Manitoba experiments in 1969 did find a response. The inconsistent S fertilizer response on deficient sites may be due to limiting amounts of moisture and other nutrients, or due to sulphate rich layers below the sample zone but within rooting depth, or due to S mineralization from organic matter. For example, many studies have found a significant interaction between N and S. Good yield response to either nutrient requires adequate supply of the other. An example of the N-S interaction is shown in Figure from research conducted by the AAFC Melfort, SK Research Centre in 1999. The need for balanced N and S nutrition is obvious without S, additions of N fertilizer lowered yield.

Field and controlled environment studies in southern Alberta found that rapeseed could utilize sulphate from a depth of 54 to 72 cm (21 to 28"). Subsoil sulphate salt layers are common in Brown, Dark Brown and Black soils on the prairies, and could affect the yield response to S fertilizer on fields testing low for S in the surface soil.

In contrast, S responses have been reported on fields with adequate soil test S levels. Recent research has measured high variability of sulphate contents across farm fields, which creates difficulty in obtaining representative samples. For example, extreme sulphate variability on a 24 ha (60 ac) solonetzic field near Stettler, AB in 1994. Composite samples were taken from each 0.5 ha (acre) and tested separately. The average worked out to be 1,076 kg sulphate/ ha (960 lb sulphate/ac) in the (60 cm (2') depth, which is excessive from the soil test standpoint. However, the average value is heavily skewed by the few samples with extremely high sulphate values (maximum 21,280 kg sulphate/ha (19,000 lb sulphate/ac). A better indicator of the most typical value for the field would be the value class that occurs most frequently (the mode). In this example, the mode was just 9 kg sulphate (20 lb sulphate), which is deficient. Therefore, the majority of this field would likely respond to S fertilizer. This example illustrates that single composite soil samples from fields with high S variability can be difficult to interpret. A soil testing deficient for S is likely deficient (unless underlain by a subsoil sulphate salt layer), while soils testing medium or high for S may have deficient areas that are skewed by areas with excessive S.

SULPHUR SUPPLY FROM THE SOIL

The organic portion of the sulphur cycle in soil is closely tied to N due to their association in protein. Like N, the main S reserve in soil is in organic matter. Although there is considerable variability in the relative proportions of carbon, N and S (C:N:S) in soil organic matter, the ratios are quite similar for each soil group. In a study of Saskatchewan farm soils, the C:N:S ratio ranged from 58:6:1 in Brown soils, to 63:7:1 in Dark Brown, 83:8:1 in Black, 100:8:1 in Gray Black, and 129:11:1 in Gray soils.

Many sulphur transformations in soil are analogous to N each undergoes mineralization from organic matter, immobilization, oxidation and reduction of inorganic compounds. The soil S cycle is illustrated below in Figure.

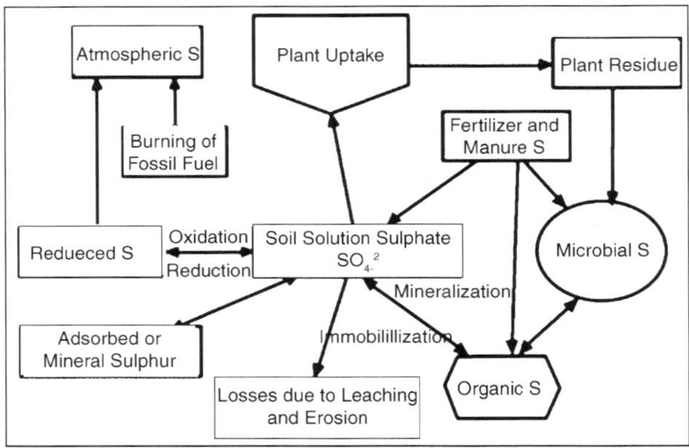

Fig. Soil Sulphur Cycle

A key component of the soil S cycle for plant growth is the mineralization path. Soil organic matter and plant residues are decomposed by soil microbes, releasing sulphate. Like N, the S mineralization rate is quite slow, and cannot match the uptake rate of growing plants. Also like N, the sulphate amounts released from residues will depend on the S content. When plant residues contain more than about 0.15% S (C:S ratio about 300:1), there will be a net release of sulphate through mineralization. Below 0.15% S, decomposition is slower and there will be immobilization of soil sulphate by soil microbes. The ability of soil to mineralize sulphate from organic matter has been found to be independent of the total amount of C, N or S, and of the C:N or N:S ratios in soils.

However, research has found that the initial amounts of sulphate mineralized from soil is closely correlated with the initial amounts of N mineralized in short-term incubation.

Another important aspect of the soil S cycle is the oxidation path. In soils, sulphides, elemental sulphur and thiosulphate can be oxidized to sulphate by various soil microbes, but the main actors are bacteria from the genus Thiobacillus. The oxidation of these inorganic S compounds produces considerable sulphuric acid. Sulphur oxidizing bacteria are most active under warm, moist, well aerated conditions. It is the oxidizing ability of these bacteria that permits the agricultural use of elemental S for crop growth.

Although S reduction is shown in the soil S cycle diagram, it generally is not significant in aerated agricultural soils. In flooded soils, sulphate can be reduced by soil microbes to sulphides in a process analogous to denitrification. However, soil microbes will utilize nitrate, iron and manganese compounds before reducing sulphate.

In many western Canadian soils, there is a subsoil salt (gypsum) and/or lime (calcium carbonate) layer. This subsoil layer contains considerable sulphate, often as coprecipitates with lime. Although this subsoil sulphate solubility is reduced, it still can contribute to plant needs if it exists within the rooting zone. However, the length of time that canola grows in S-deficient topsoil before rooting to the subsoil S will affect the yield response to fertilizer S. Also, the depth to subsoil S tends to vary greatly across the field. Total S amounts (organic and sulphate) generally increase from upper to lower slope positions.

SULPHUR FERTILIZER MANAGEMENT

The optimal method and timing of S fertilizer depends on the fertilizer form sulphate versus elemental forms.

Sulphate S

Sulphate fertilizers are highly soluble and will move easily with water in the soil. Ammonium sulphate (21-0-0-24) is a common sulphate-based fertilizer. Liquid ammonium thiosulphate (12-0-0-26) is used less frequently, and requires a short time period for oxidation to sulphate. Highest fertilizer use efficiency generally results when sulphate fertilizer is placed near roots for easy access, and just before the period of plant uptake. Under dry spring conditions, broadcast sulphate fertilizer can be stranded and result in poor uptake. However, under such dry conditions, canola germination and establishment will also be severely affected. Under average to good moisture conditions, sulphate fertilizer can be broadcast-incorporated in the spring with good results. On sandy soils, sulphate leaching can occur during wet periods. Therefore, apply fertilizer just prior to crop needs.

Although sulphate fertilization just prior to or at seeding is best, post-seeding applications can be effective. Research has found that sulphate fertilizer can be soil applied up to the rosette stage with good response provided there is sufficient rain to wash the fertilizer into the root zone before bolting. Foliar sulphate fertilizer solutions can be applied up to the bolting stage with moderate effectiveness, and can serve as a rescue treatment for deficient fields that were not adequately fertilized at or before seeding. However, the best yield response to sulphate S fertilizer is at or before seeding as illustrated in Figure from six site-years of recent research at AAFC Melfort, SK Research Centre.

In the above research, seed row placement of 30 kg S/ha (27 lb S/ac) increased yield less than side banding or preseed incorporation, probably due to seedling toxicity. This agrees with other research, indicating that only limited amounts of sulphate S fertilizer can be safely applied near the seed. The safe amount will vary with the degree of seedbed utilization, moisture conditions, soil type and amounts of other fertilizer nutrients. Until research has determined the safe amount of seed row S fertilizer under various combinations of the above factors, limit seed-placed ammonium sulphate. Include N amounts in sulphate fertilizer when calculating total N fertilizer amounts to be safely seed placed.

Elemental Sulphur

Sulphur fertilizer containing elemental sulphur must be managed differently than sulphate-based fertilizer to achieve good efficacy. Elemental sulphur has advantages of a ready supply in western Canada, low production and transportation costs, and fewer drill fill operations due to high analysis. However, elemental sulphur has a significant disadvantage availability is delayed until soil bacteria oxidize it into the sulphate form. The conversion rate from elemental sulphur to sulphate depends on the particle size, the degree of dispersion in the soil, and the growing conditions for the bacteria (moisture, temperature).

Common elemental sulphur fertilizers are formulated as granules or pastilles (split pea shape) for ease of shipping and handling, each consisting of thousands of individual particles. The surface area of these individual particles is the access where the soil bacteria feed converting the elemental sulphur to sulphate. Small particles have the largest surface area and the fastest oxidation rate. Under western Canadian conditions, research indicates that particles less than 150 microns in size will convert quickly if well mixed with soil. Some elemental sulphur fertilizers have particles consistently smaller than 150 microns such as Sulpher 95. Other products consist of a mixture of particles ranging from smaller to larger than 150 microns such as Tiger 90CR In addition, some products such as iger 90CR contain bentonite clay that swells when wet and helps to break the granule into the small particles. Granules that break down readily and completely will allow quicker oxidation and sulphate availability.

Research in western Canada has found that elemental sulphur granules break down the greatest when applied to the soil surface and exposed to rain/

snow and frost. Subsequent tillage will then further disperse the degraded granule. In contrast, band and seed row placement, or immediate incorporation following broadcasting, will reduce the granule dispersion and the oxidation rate. The data taken from a recent lab experiment by the Agronomy Unit of Alberta Agriculture, Food and Rural Development.

The strong influence of elemental sulphur placement and timing on sulphate release subsequently affects canola yield. This effect is evident in Table which is based on research by AAFC at Melfort, SK. The effect of sulphur fertilizer type, placement, and application time on canola and wheat yield was studied. All S was applied at 20 kg S/ha (18 lb S/ac) in the first year only to canola and wheat plots. To measure residual response over three years, canola was seeded in the second year onto wheat plots grown with S treatments in the first year. No S was applied in year two. Canola was seeded in the third year onto wheat plots from year two, again without S fertilizer. Ammonium sulphate and ammonium thiosulphate corrected S deficiencies in the application year and provided residual response over the next two years. Broadcast elemental S fertilizers did not relieve S deficiency in the year of application, but broadcast Tiger 90 did provide a response equivalent to ammonium sulphate two years later. Pre-seed banded or seed-placed elemental S did not correct S deficiency, even two years later. Continuing research shows that fall broadcast elemental S fertilizer left on the surface until incorporation next spring will improve the oxidation and canola yield response compared to spring broadcast. However, this practise should still be initiated two years ahead of the canola crop to ensure consistent response.

Another factor that influences the oxidation rate is previous use of elemental sulphur in the field. Exposure to elemental sulphur in the past has been shown to increase oxidation rates of subsequent applications, probably due to stimulation of the S-oxidizing bacterial population.

However, repeated applications of elemental S fertilizer can negatively impact agricultural soil. Saskatchewan research on Gray Wooded soil found that repeated elemental sulphur application decreased soil pH, organic C, and microbial biomass. Negative effects on soil enzymes involved with nutrient transformations were also found. Although short term crop response to elemental sulphur fertilizer is usually less than with sulphate sources, there may be residual benefits of elemental sulphur due to the slow release. While a residual benefit has been found with forages, conflicting results have been found with canola. Sulphate generated from elemental S after the period of annual crop uptake (May to August) may not benefit the soil S status in subsequent years since it is susceptible to leaching during snowmelt. Conditions most likely to exhibit residual benefit from elemental S fertilizer would be combinations of sandy, low organic matter soil, sloping topography, and high rainfall episodes after fertilizer application but before major uptake has occurred.

NITROGEN:SULPHUR RATIO

A proper N:S balance is important for canola production. When N is in excess (high N:S ratio), there is insufficient S to combine with the N to make protein, and thus nonprotein N accumulates. A useful guideline is to add N and S fertilizer in a 7:1 ratio, which is approximately the ratio needed by the canola plant. There has been research into using the N:S ratio during tissue testing to determine S status. However, the N:S ratio of canola tissue has not proved reliable for predicting S status. The N:S ratio only indicates the relative proportions of N and S in the plant, and does not indicate their actual magnitudes. Therefore, if canola tissue tests show an optimal ratio of 7:1, there are three possibilities: both N and S levels are optimal, excessive, or deficient.

At the rosette stage, tissue testing canola for S status should include several criteria to improve the reliability:

- % S greater than 0.25%
- N:S ratio of 10 or less
- A sulphate:total S ratio (as indicated by hydriodic acid reducible S:total S) greater than 0.38

MICRONUTRIENTS

Micronutrients are those nutrients required in extremely small quantities (less than 100 ppm in plant dry weight). Unfortunately, the basic functions of micronutrients are less understood than macronutrients. Also, there is very limited knowledge about the forms and mechanisms of micronutrient transport in the xylem and phloem.

Micronutrient deficiencies in canola are much less common than macronutrient deficiencies. However, canola yields can be severely depressed when micronutrient deficiency occurs. This section will review the various micronutrients and canola responses.

COPPER (CU)

Knowledge about copper fertility of western Canadian soils has increased over the past two decades. Previously, Cu deficiency was thought to be limited to organic soils. The recent research has identified that Black, transitional Gray-Black and Dark Brown soils may be Cu deficient for cereal production.

Although copper deficiency and fertilizer response has been documented with cereals under field conditions, canola has not been shown to display deficiency symptoms or respond to Cu fertilizer.

Role of Copper in the Canola Plant

Copper is a transition element that forms stable complexes in the plant and soil, and is capable of electron transfer (energy processes). Copper role in plant functions is mainly as a reactive constituent of enzymes that catalyze oxidation-reduction reactions.

Some examples of Cu containing enzymes include:

- Plastocyanin (needed for energy capture through photosynthesis)
- Superoxide dismutase (needed for detoxification of oxygen radicals)
- Many different types of oxidases (enzymes that degrade or change compounds together with oxygen)

One of the phenol oxidases is involved with lignin synthesis.

Due to coppers role in photosynthesis, deficiency leads to low carbohydrates levels, at least during the vegetative stage. The low carbohydrate content in Cu deficient plants contributes to impaired pollen formation and fertilization. The reduced lignification in Cu deficient plants also affects pollen fertility since lignification of anthers is needed to release pollen.

Copper Supply from the Soil and Uptake by Canola

Copper is a metallic nutrient that originates from minerals in the soil. The total Cu content in western Canada soils usually falls in the range of 5 to 50 ppm. Approximately 1/2 to 1/4 of the total Cu exists within minerals and is unavailable to plants. Copper associated with oxides and organic matter has been found to be an important source of plant available Cu, probably by replenishment of dissolved and exchangeable Cu. The oxide and organic fractions increase with the clay content, which explains why Cu deficiency is more likely on sandier textures. Exchangeable Cu ranges from 0.1 to 10% of total soil Cu. Only very small amounts of Cu exist as soluble Cu+2 in the soil water. Research on the prairies has found that DTPA extractable Cu is highly variable across cultivated and native fields. This means that larger numbers of soil samples are needed to obtain a precise estimate of the true soil average.

Very little information exists on the mechanisms of Cu uptake by plant roots. The driving force for Cu uptake is the electrical chemical gradient across the root cell membranes. Since free Cu levels inside the cell are kept low to avoid harmful reactions, and the membranes have a large negative potential, this creates a large force for Cu uptake. Therefore, there is no need for active Cu uptake systems. It has been suggested that Ca channels likely also allow passage of other ions such as Cu+2. Copper remobilization is much higher in old leaves and is related to N remobilization. Despite the intermediate mobility of Cu, deficiency symptoms in sensitive crops during the vegetative stage first appear in new growth. However, canola does not display strong Cu deficiency symptoms.

Pot experiments with extreme Cu deficiency have reported canola symptoms of:

- Interveinal chlorosis shortly after emergence
- Larger than normal leaves
- Wilting leaves
- Delayed flowering with a shortened flowering stem

Evidence suggests that canola growth may be affected by imbalances between Cu, molybdenum (Mo) and manganese (Mn) levels.

Manganese:copper ratios (DTPA extractable) greater than 15 may result in a Cu deficiency. Molybdenum also may antagonize Cu. However, the Mo antagonism is in turn affected by S levels. Sulphur additions were found to lower Mo contents in canola plants, reducing Mo antagonism with Cu, and Cu deficiency was alleviated without adding Cu fertilizer.

Canola Response to Copper Fertilization

Although numerous field experiments have shown yield responses in cereals to Cu on deficient soils in western Canada, no positive reports in canola have been published. Two growth chamber experiments using extremely deficient organic soil did find a response to Cu fertilizer. Therefore, organic soils are likely the only soils in western Canada that may show a Cu response in canola. A compilation of research data from Saskatchewan and Alberta on mineral soil suggests that 0.30 ppm diethylenetriaminepentaacetate (DTPA) extractable Cu may be the critical level for canola. Since the critical Cu level is much lower for canola than cereals, Cu fertilization programmes should focus on application to the cereal rotation phases.

MANGANESE (MN)

Manganese is a metallic micronutrient that is occasionally deficient in western Canadian organic, high pH soils. Although oats can be affected by Mn deficiency in cold organic soils (gray speck of oats disease), there have been no documented problems with canola. Weathering of manganese containing soil minerals is the source of plant available Mn. The main Mn form that exists in soil solution or adsorbed to soil colloids is Mn+2, which is also the form absorbed by roots. Manganese availability decreases when pH increases above 6.2 in many soils. Low temperature and high organic matter can also decrease Mn availability. Rhizosphere microbes play a role in Mn availability by either oxidation or reduction. These microbes can either increase or decrease Mn availability. The rhizosphere acidification by canola roots likely increases Mn availability and makes this crop relatively tolerant of low soil Mn. Plant Mn status is affected by Cu levels. High Mn:Cu ratios above 15 may lead to Cu deficiency while ratios below 1 may lead to Mn deficiency. Potash (KCl) has been shown to enhance Mn uptake by several crops. In some crops, seed Mn content has been shown to be important for initial Mn nutrition and plant growth, as well as disease resistance.

BIOLOGICAL AND CHEMICAL ASPECTS OF SOIL PRODUCTIVITY

Organic matter, microscopic and macroscopic organisms (*e.g.*, fungal hyphae and invertebrates), detritus from fungi and animals, and bacteria, and biological exudates, all assist in stabilising soil structure. The role of each part of the biomass differs according to its size. Broadly, large aggregates greater than 250 m m diameter (macro-aggregates), are stabilised by their inherent physical structure, wetting and drying cycles, and organic matter. Micro-aggregates (< 250 m m) are stabilised by live or dead roots, fungi, invertebrates and micro-organisms. The populations of soil organisms of all sizes are linked functionally through their roles in the degradation of various forms of organic material. The latter includes live and dead plant material and other live or dead organisms. The animals such as nematodes and some fungi feed directly on live plants while other fungi and bacteria feed predominantly on litter. Earthworms and other large invertebrates create, and inhabit, burrows and pores, and are very mobile.

The most notable of these are termites, which are divided into three groups according to the structure of their nests: those that build mounds:

- Above ground.
- On the soil surface.
- Below ground.

Small arthropods, microfauna and fungi live mostly in larger voids and in association with roots. Foster reviewed the location of the various types of soil-dwelling organisms and found that fungi, which constitute about 80 per cent of the biomass in many soils, tend to be restricted to the rhizosphere of roots, to larger pores between aggregates and to the surface of aggregates. Bacteria, by contrast, are found on roots in the rhizosphere, in small colonies in the larger micropores, within aggregates and on and within cell debris.

ORGANIC MATTER IN PLANTS AND ANIMALS

Both plants and animals provide inputs of organic matter to soils. Once within the soil organic residues can be distinguished on the basis of their chemical structure (*e.g.*, old lignified humic substances that degrade slowly), by their source (plant or animal) or by location. The standing crop of litter in semi-arid grasslands is usually more than 3 t/ha and in temperate dry steppe may exceed 11 t/ha. There has been much debate about the relative contents of organic matter in tropical and temperate soils. Within those wet-and-dry climates that have hot summers assisting rapid decomposition, there is no evidence of inherently lower levels of organic matter in the tropics than in comparable temperate regions.

Kowal and Kassam and Juo and Payne review the role of organic matter in tropical soils. Here, it is sufficient simply to state that organic matter has various interrelated effects on soil fertility. In particular it should be noted that both chemical and physical effects are of relatively great importance in the soils of the semi-arid tropics because these generally have low cation exchange capacity.

The relative importance of litter (crop residue) and manure as inputs of organic matter, varies between cropping systems and spatially within a system. The flow of litter, manure and by-products (such as dung cake for fuel) in Indian villages practising approximately one-third single cropping and two-thirds double cropping. Here, most of the above-ground crop residue is fed to animals, but an equal amount of below-ground crop material enters the soil organic matter pool. As much dung is distributed naturally to grassland and forest than is recycled to the crop.

The pattern of flow of manure and litter to the cropping soil falls within the wide range of practice common in wet-and-dry climates. Generally, all the root material (say, 40 per cent of total crop growth) and 10-30 per cent of top material may be recycled in an annual cropping system. Where alley cropping and agroforestry are practised, values are more variable, but possible inputs could be very significant where the trees, from which the litter is taken, are grown away from the annual crops.

If, say, two-thirds of the leaves from leguminous trees are harvested annually, litter values will be substantially higher and the material of better quality than the leaf and stem residue from an annual crop likely to be recycled in the field. Some qualifications, however, should be made. Tree root material is not available for decomposition in the crop field unless it is spatially overlapping (*e.g.*, as an intercrop), in which case the trees will compete with the crop for soil nutrients, water, light and space.

The proportion of animal and human manure used on cropland is more variable. Some farmers have developed stable systems which strongly emphasise the use of animal manure on crops. For example, Norman *et al.* describe how farmers in northern Nigeria managed to apply 4 t/ha of manure to their heavily-cropped crop land though they had only 3 cows each. Many other farmers do not ensure adequate recycling, either because they are more concerned with livestock management or they do not know the importance of maintaining a 'zero nutrient budget' to replace nutrients removed by the crop. For example, Norman *et al* also describe farmers with 10 cows each, who applied only 1.9 t manure/ha to their crops.

Within a cropping system, manuring practice varies with location. There is transference towards the centre of the system. On traditional farms, the area near the household or village is highly fertilised with human and animal manure while more distant fields receive little or no organic matter. Fussell describes such a traditional 'ring' farming system in semi-arid west Africa. Here, if houses are thatched, the village needs rebuilding or moving every 2 to 4 years. Moving takes advantage of the fertility gradient. Where the huts are not moved, the fertility gradient becomes steeper with time. Rather than trying to even out fertility by labour-intensive transport of manure, farmers vary the cropping of the fields. Continuous cropping of millet is sustainable close to the hut or village where there is plenty of human and animal manure but crop rotations are essential at the periphery.

Closed systems, which rely on maintenance of fertility by recycling organic matter, are not sustainable. Nitrogen levels may be maintained or even enhanced by using legume crops or pastures in rotation but even then further inputs of nutrients are required to replace those removed by the crop. Fussell describes the inevitable need to use inorganic nutrients or organic nutrients from off-farm sources such as wastes from cities to complement farm organic matter in the African Sahel:

"Although the importance of organic manure in increasing cereal yields is well recognised, it is unlikely that there is sufficient available, on a regional basis, to sustain yields without the use of additional chemical fertilizers. For example in the Western African savannah 10 t/ha/year of manure are required for sustainable cropping of millet or sorghum.....while in northern Nigeria 2.5 t/ha/year are sufficient to maintain yields in most areas." This four-fold difference largely reflects the integration of livestock with cropping in Nigeria which results in relatively high levels of soil organic matter.

Farmers elsewhere, for example in Ghana, give lower priority to crop fertility, which lowers soil organic matter levels and leads to greater nitrogen loss through water erosion when there is little litter cover on the ground. Irrespective of these differences, however "It is doubtful if either of these levels of application can be achieved, as... estimates cattle and sheep manure outputs in the northern savannah zone of Nigeria are estimated to be only 1.4 and 0.25 t dry matter per head per annum respectively." Off-farm sources of manure, organic matter generally (*e.g.*, industrial by-products) and soil ameliorants will become increasingly important. Some, such as sewage from cities, are being researched. Their use in developing countries seems likely to increase, particularly as labour for transporting these materials is cheap relative to the cost of other nutrients such as manufactured inorganic fertilizers. The treatment and efficacy of various off-farm sources of organic matter are outside the scope of this *Bulletin*. It suffices to point out that substantial amounts of material are available.

The rate of breakdown of litter and other organic materials both determines and depends on the populations and activities of organisms in the soil. It also determines the extent to which minerals taken up by a crop are released from its organic residues and made available as inorganic ions to the subsequent crop. There are numerous mathematical models of organic matter degradation. Almost all assume that the rate of degradation decreases with time, as the more soluble, digestible and accessible material is selectively and progressively degraded by soil organisms. This selectivity is often arbitrarily described by recognising several 'pools' of organic matter and, with time, 'passing' material from labile to less-labile compartments. In practice its physical and chemical composition changes gradually with time, as does its physical binding with soil particles and its relation to the predators that feed on it.

The rate of decomposition of leaf litter depends on its environment, particularly on soil temperature and soil water. Both these affect the physical breakdown of the litter and determine the population and activity of soil animals and fungi that feed on it. Decomposition also varies with plant type and age of litter, being slower for heavily lignified material. The specific properties of litter from different species, and the generally exponential form of litter decay (the rate of decomposition slowing with time) lead to values that suggest half-lives of litter ranging from 1 to about 10 years. Because of the importance of temperature in determining decomposition, rates decline from tropical to temperate locations within wet-and-dry climates.

Plant species, water and temperature are not the only factors affecting the speed of litter decomposition and mineral cycling. Management plays a role.

There are four ways for managing crop residues:

- Stubble mulch in which residues are left standing;
- Surface mulch, where above-ground residues are cut and left on the top of the soil after harvest;
- Incorporation by ploughing; and
- Cut-and-carry, in which surface residue is removed and (if not used for livestock or thatching, etc.) returned as a surface mulch about planting time for the subsequent crop; this is usually combined with ploughing of below-ground residues.

Two of the above treatments involve the fragmenting of root residues. All of them affect a range of soil properties. The degree of contact between the residue and the soil (and its organisms) and other unexplained factors affect the rate of decomposition.

The rates of release of minerals from organic matter, particularly litter, are a significant element in nutrient cycling and crop productivity in semi-arid environments. Such climates have a rainy season in which rainfall exceeds evaporation, so there is a risk of mineral loss through run-off and mineral loss and acidification by leaching. During the dry season there is potential loss of nitrogen as ammonia. This suggests the need to manage litter so most decomposition takes place when the subsequent crop is growing rapidly and is able to take up the nutrients released. It is wise at the same time to maintain ground cover if possible. Manipulation of litter requires labour, which may already be in maximum demand prior to sowing.

IMPORTANCE OF SOIL BIOLOGY

The activities of the wide range of organisms in soil play a pivotal role in both natural and managed ecosystems. Their processes of organic matter breakdown contribute to the soil's health – its stability, permeability, ability to retain nutrients and make them available for plant uptake.

Soil biological, physical and chemical processes are interrelated and all contribute to plant productivity. The level of soil biological activity is therefore affected by the soil type, but it also depends on the management practices used, particularly the management of organic matter, especially carbon. Changes that are made to the chemical and physical environment in soil will therefore influence the biological processes and subsequently the contribution they make to the soil's fertility overall.

ISSUES ASSOCIATED WITH SOIL BIOLOGY

Knowledge of soil biological processes can support decision making aimed at achieving sustainable use of agricultural land. Soil biology is a complex field, however, and research continues to uncover new facts concerning the organisms themselves, their processes and factors that affect them. Additionally, management of soil biological processes is difficult to do precisely due to the differing parameters of each individual situation, such as soil type and land use. A certain land management practice may also affect one group of organisms, but not others. Consideration of the mass of microorganisms as a whole (microbial biomass) is therefore not sufficient for a complete interpretation of the effects of land management on soil biological fertility.

Measurement of soil health in terms of biological fertility is also a complex and at present, relatively expensive process. More attention has been usually given to the management of the soil chemical and physical environments. As a result, a host of inexpensive and simple tests are available to land managers. Incorporating soil biological processes into farming systems will require a more holistic and longer term approach to agricultural land management.

BENEFITS OF MANAGING SOIL BIOLOGY

Agricultural land management practices alter aspects of soil chemical and physical fertility with consequences for soil biological processes and vice versa. Both fauna and microorganisms contribute significantly to chemical transformations in the soil and influence their physical surroundings to various degrees. Organisms on and around plant roots have major influences on plant nutrient availability and some form specific associations with legumes, which greatly influence the C:N ratio of plant residues in soil. Associations between agricultural plants and fungi known as arbuscular mycorrhizas have the potential to increase the efficiency of use of phosphorus in agricultural ecosystems as well as improve soil structure.

Current research includes understanding some of the undesirable members of the biological population, such as root pathogens. The role of mycorrhizal fungi in preventing and reducing the effects of salinity is also being investigated. Stubble management is another area of research.

SOIL COMPONENTS AND BIOLOGICAL PROCESSES

Soil is a dynamic medium made up of minerals, organic matter, water, air and living creatures including bacteria and earthworms. It was formed and is forever changing due to five major physical factors; the parent material, time, the climate, the organisms present and the topography. The way in which we manage soil is another major factor influencing the character of the soil. Note that the use of the term mineral describes the occurrence of crystals of quartz, mica, etc., and is not used to indicate plant nutrient elements. Soil is not uniform on the surface, and it is not uniform with depth. It has layers or horizons.

If you dig a pit to about 2 metres deep, you can observe these soil horizons. There is topsoil, or A horizon, where all the organic matter accumulates and is consequently the most fertile horizon. Some soils do not have an A horizon because it has been eroded away, for example, after a wind erosion event. Needless to say, these soils are often very infertile. Below the A horizon is the B horizon, or the subsoil.

This is less fertile than the A horizon and usually has more clay, iron and aluminium compounds, as water has leached the finer particles down the profile. Below the subsoil is the parent material, the rock from which the soil has formed. Physical, chemical and biological weathering of this rock eventually leads to the formation of soil. Valley floor and plateau soils tend to be deeper than those on the slopes due to erosion removing the soil from the slopes.

There is more time for soils to form in valleys and plateaux. Organisms add nutrients to the soil, aid in improving soil structure and help stabilise the soil against erosion, once again changing the soil.

SOIL ORGANIC MATTER

Organic matter is anything that contains carbon compounds that were formed by living organisms. Examples include lawn clippings, leaves, stems, branches, moss, algae, any animal parts, manure, sawdust, insects, earthworms and microbes.

There are 3 main components of organic matter in soils:

1. Dead and decaying forms of organic material - mainly dead plant parts
2. Living plant parts - mostly roots
3. Living microbes and soil animals

The largest component of organic matter is the dead and decaying materials. For example, it can constitute approximately 85 per cent of organic matter in soil. Living roots can make up about another 10 per cent and the microbes and soil animals make up the remainder. The exact amounts vary with soil types and plant community.

HUMUS

Organic matter that is so decomposed that it can no longer be recognised as individual components is known as humus. The highly complex compounds that make up humus are able to resist further decomposition, and therefore accumulate in the soil.

Impact of Incorporating Organic Matter Into the soil

Incorporating organic matter into soil can alter the physical, chemical and biological balances in the soil.

It can change the:

- Amount of nitrogen that is available to plants
- Amount of other nutrients available
- Way the soil holds together (soil aggregation)
- Number and type of organisms present in the soil

All of these changes are related to the way the organic matter is decomposed when it is incorporated into soil and to the particular type of organic matter present.

Nutrients taken up by plants in natural environments are derived largely from decomposition processes. During organic matter decomposition, bacteria and fungi can use some of the energy or nutrients released by that process for their own growth. For example, during the break down of a protein, microbes will have access to the carbon, nitrogen and sulphur that become available for their own physiological processes and cell structure. If there are nutrients surplus to the microbes' needs, they will be available for other soil organisms or plants to use. When microbes die, their cells are degraded and nutrients contained within them again become available to plants and other soil organisms.

Incorporating organic matter into soil can change the amount of nitrogen (and other nutrients) that is available to plants. Microbes can access nitrogen in the soil more quickly than plants. Depending on the C:N ratio of the organic matter they exploit, mineral N may be released during decomposition. However, the microbes will use some or all of the N first, or retain it in microbial tissues, rendering it temporarily unavailable for plant use. If there is insufficient N in the organic matter, most of the N may even be taken up from the soil by the microbes, leaving a lower amount for plant use; this is called 'immobilisation'.

Adding organic matter can also increase the activity of earthworms, which in turn can improve soil structure. Similarly, if the amount of fungi and bacteria associated with the breakdown of organic matter increases, soil structure will again benefit. Organic matter that is durable and less easily degraded will increase the microbial biomass over a longer period of time.

SOIL AGGREGATION

Soil aggregates are 'clumps' of soil particles that are held together by moist clay, organic matter (such as roots), by organic compounds (from bacteria and fungi) and by fungal hyphae (pronounced "highfee"). Aggregates vary in size from about 2 thousandths of a millimetre across up to about 2 millimetres across, and are made up of particles of varying sizes.

Some of these particles fit closely together and some do not and this creates spaces of many different sizes in the soil. These spaces, or pores, within and between soil aggregates are essential for storing air and water, microbes, nutrients and organic matter. Soils with many aggregates are called "well-aggregated". Such soils are more stable and less susceptible to erosion.

FORMATION OF A SOIL AGGREGATE BY MICROBES

Bacteria

There are two ways that bacteria could be involved in soil aggregation. One way is by producing organic compounds called polysaccharides. Bacterial polysaccharides are more stable than plant polysaccharides, resisting decomposition long enough to be involved in holding soil particles together in aggregates. The other way bacteria are involved in soil aggregation is by developing a small electrostatic charge that attracts the electrostatic charge on clay surfaces, bringing together small aggregates of soil.

Fungal Hyphae

Fungi grow in long, threadlike structures, called hyphae. The amount of aggregation in the soil has been found to relate to the length of fungal hyphae in the soil. Fungi help to form aggregates in the soil by enmeshing soil particles with their hyphae and forming cross-links between soil particles. Mycorrhizal fungi and fungi that colonise fresh organic matter are believed to be the most important for assisting with stabilisation of soil particles into aggregates.

DEGRADATION OF PESTICIDES

Chemicals, such as pesticides, added to the soil are broken down in a similar way to organic matter. They are oxidised by the microorganisms that produce the appropriate enzymes. Pesticides generally have very complex structures and most are converted to several intermediate organic compounds before finally being converted to carbon dioxide and water.

When these kinds of chemicals are degraded metabolically, the microorganisms involved gain energy and carbon from the oxidation of the chemical. This allows the microorganisms to grow and multiply, and the size of the microbial population may increase as more of the chemical is degraded. Sometimes, microorganisms capable of degrading a particular chemical grow and multiply until there is a higher proportion of these microorganisms in the soil.

When a community of microorganisms in the soil adapts in this way, the chemical can be broken down more rapidly than it would have previously. This can be a problem when rapid degradation of a pesticide may reduce its effectiveness against pests. Some chemicals added to the soil are degraded by the microorganisms without any particular benefit to the organism such as nutrients or energy supply. The population of microorganisms remains constant while the concentration of the chemical in soil decreases. The different modes of interaction between synthetic chemicals and microorganisms are a result of the enormous diversity of organisms in soil.

Some Chemicals Easier to Degrade than Others

The ease with which a chemical is degraded depends on the complexity of its structure. More complex structures degrade more slowly either because fewer microorganisms in the soil produce enzymes capable of degrading them or because of the inaccessibility to microbes.

THE NITROGEN (N) CYCLE

The nitrogen cycle describes how the different forms of nitrogen in the air, soil, water and living organisms are interconnected. It is described as a cycle because the nitrogen is never lost completely; it just changes form and is held in different places. There are 5 major processes involved:

- Ammonification
- Nitrification
- Denitrification
- Nitrogen Fixation
- Nitrogen Immobilisation.

Ammonification

Ammonification is the conversion of organic forms of nitrogen (for example, nitrogen in proteins in dead plants and soil animals) to ammonium. Ammonium is an inorganic form of nitrogen that has the symbol NH_4^+. A wide range of soil organisms carry out the ammonification process. Many different types of bacteria and fungi are involved. The organisms get carbon and energy from the breakdown of organic matter, while nitrogen is released at the same time.

Nitrification

Nitrification is the process where ammonium is converted to nitrite (NO_2^-) and then to nitrate (NO_3^-). The organisms involved in nitrification are a relatively small number of species in soil.

They are autotrophic, which means they get energy from the chemical transformation of NH_4^+ to NO_2^-, while any needed carbon is obtained from CO_2. One group of organisms converts ammonium to nitrite and another group of organisms converts nitrite to nitrate.

Denitrification

Denitrification is the process where oxides of nitrogen (*e.g.,,* nitrate and nitrite) are converted into gaseous nitrogen and are removed from the soil system. Examples are the gas nitrous oxide (NO_2), and nitrogen gas (N_2), which is the form of nitrogen most common in the atmosphere. Many species of microorganisms are capable of denitrification of soil. Denitrification occurs mainly when there is little or no oxygen in the soil, such as when the soil is waterlogged. However,. if the waterlogging is only temporary, the denitrification process will stop when the soil dries.

Nitrogen Fixation

Nitrogen fixation is the conversion of nitrogen gas (N_2) to ammonium (NH_4^+)- either by free living bacteria in soil or water, or by bacteria in symbiotic association with plants (eg legume symbiosis).

Nitrogen Immobilisation

This is the process whereby nitrogen is incorporated into the microbial cells and 'tied-up' in the 'microbial pool' of nitrogen. Immobilisation occurs in parallel with ammo-nification.

THE SOIL HABITAT

Organisms Located in Soil

Most organisms are found in the top layers of soil, usually the top 2-3 centimetres, since, this is typically where most of the organic matter is. Organisms do occur to depths of several kilometres below the soil surface, but the types of organisms that occur this far down are not the same as those close to the surface. The organisms in soil are often commonly found close to root surfaces in the rhizosphere, within living and dead roots, on soil particles, or amongst aggregates of soil particles.

Earthworms and other soil animals are able to move through most of the top layers of soil. Fungi can form a mat of hyphae, which can extend centimetres or even metres through the soil. They can also form a network of hyphae inside soil aggregates. Bacteria tend to accumulate inside soil aggregates because they are less likely to be eaten by soil animals such as protozoa and mites in this environment. Bacteria can be carried down further into the soil in water that is percolating downwards, but generally they do not move far. Soils that are clayey often have more bacteria than sandy soils because the clay creates lots of small pores (spaces) which offer protection for bacteria. Sandy soils with fewer aggregates and small pores are less suitable habitats for bacteria and fungi unless a large amount of organic matter is added to the soil.

Characteristics of the Rhizosphere

The rhizosphere is the region of soil that is immediately adjacent to and affected by plant roots. The rhizosphere is a very dynamic environment where plants, soil, microorganisms, nutrients and water meet and interact. The rhizosphere differs from the bulk soil because of the activities of plant roots and their effect on soil organisms.

A major characteristic of the rhizosphere is the release of organic compounds into the soil by plant roots. These compounds, called exudates, make the environment of the rhizosphere very different from the environment in the bulk soil. The exudates increase the availability of nutrients in the rhizosphere, and also provide a carbon source for heterotrophic microorganisms. The exudates cause the number of microorganisms to be far greater in the rhizosphere than in the bulk soil. Their presence attracts larger soil organisms that feed on microorganisms and the concentration of organisms in the rhizosphere can be up to 500 times higher than in the bulk soil. However, their growth and reproduction is even higher when grasing by predators is taken into account. This grasing helps release nutrients in microbial pool back into the soil. Thus the turnover of nutrients can be more important than the mount of microbes present at any point in time.

Another characteristic of the rhizosphere is the uptake of water and nutrients by plants. Water is drawn from the surrounding soil towards the roots. The balance between the movement of water and nutrients towards the roots and their removal from the soil by roots means that their concentration in the rhizosphere is usually very different from what it is in the bulk soil. This can affect microbial growth and activity.

Rhizoplane

The rhizoplane is the surface of plant roots in the soil. The rhizoplane is the site of water and nutrient uptake and the release of exudates into the soil. Like the rhizosphere, the rhizoplane is a constantly changing environment. As roots grow they cast off dead cells and navigate around soil particles, making the rhizoplane highly irregular, blurring the dividing line between the root surface and the soil.

Organisms in the Rhizosphere Influence Plant Roots

- Organisms in the rhizosphere can affect the plant roots by altering the movement of carbon compounds from roots to shoots. This alteration occurs when organisms compete with root cells for carbon that is fixed by photosynthesis.
- The burrows of earthworms in soil provide an easy route for plants roots as they grow through the soil.
- Various root microorganism associations can increase nutrient uptake by plants in nutrient poor environments, such as symbiotic (*e.g.*, mycorrhizal) and other specific (*e.g.*, nitrogen fixing) associations and rhizobia.
- Some soil organisms are pathogenic and attack living plant roots.
- Azospirillum is a bacterium that commonly lives in the rhizosphere

of grasses. Some strains of this organism produce hormones that stimulate plant growth. Other strains of Azospirillum fix atmospheric nitrogen and may or may not make it available to the grass.

Soil Organisms Dependent on each Other

The soil food web is a way of illustrating the way in which soil organisms relate to each other based on what they eat. The soil food web begins with organic matter, such as crop residues, pasture or any animal or plant material in the soil. Bacteria and fungi consume organic matter and are, in turn, consumed by nematodes, protozoa, earthworms, collembola and some mite species. Nematodes and protozoa are consumed by some mites. Mites and collembola are eaten by beetles and ants. It is important to remember that the soil is a very complex habitat, in which the diversity of organisms is greater than that of the most diverse plant or animal community. This also means that it s not easy or accurate to generalise about all soil food web processes because the extent to which they occur will be different in different soils, at different times. The same processes (eg. mineralisation) can be carried out by different groups of microorganisms.

SOIL ANIMALS

Soil animals are an extremely diverse group of organisms. Soil animals are grouped roughly according to their size, into three groups. The first group is the microfauna. These are the smallest of the soil animals ranging from 20 - 200 μm. The main soil animals in this group are protozoa. The mesofauna is the next largest group and range in size from 200 μm - 10mm. The most important animals in this group are mites, collembola (or spring tails) and nematodes. The macrofauna contain the largest soil animals such as earthworms, beetles and termites. Generally, the most common soil animals are protozoa, nematodes, mites and collembola.

Importantance of Soil Animals

Soil animals perform several functions in soil that make them a vital part of all ecosystems, including agriculture. Soil animals are involved in:

- Degradation of organic matter and mineralisation of nutrients,
- Controlling populations of pathogens,
- Improving and maintaining soil structure and
- Mixing organic matter through the soil.

Decomposition of Organic Matter by Soil Animals

Soil animals contribute directly to nutrient cycling in soil when they release mineralised nutrients in their excreta. However, most of their contributions are indirect, by:

- Grasing on the microbial biomass, which alters the rate at which organic matter breaks down

- Fragmenting organic matter and increasing its surface area for attack by microorganisms,
- Controlling the grasing pressure of nematodes on microorganisms,
- Mixing soil and organic matter and
- Introducing microorganisms onto fresh organic matter.

Soil Structure Affected by Soil Animals

Soil animals have an important role in the formation of soil structure. Soil animals improve soil structure by forming channels and pores, concentrating fine soil particles together into aggregates and by fragmenting and mixing organic matter through soil.

How Soil Animals Survive in Soil

Soil animals are found in almost every environment on earth, including Antarctica! Different species require different conditions to grow and survive but all soil animals, require sufficient carbon and nutrients, moisture, oxygen and an optimum pH and temperature. The optimum pH and temperature levels vary between species. Some organisms do not survive dry or very cold conditions, but they may leave eggs in the soil that hatch when conditions become more favourable. Other soil animals remain in the soil in an inactive state and become active again when conditions become favourable. Larger soil animals such as earthworms may live deeper in the soil when there are unfavourable conditions near the soil surface.

SOIL BIOLOGICAL PROPERTIES

SOIL BIOTA

The soil contains a vast array of life forms ranging from submicroscopic (the viruses), to earthworms, to large burrowing animals such as gophers and ground squirrels. Microscopic life forms in the soil are generally called the "soil microflora" (though strictly speaking, not all are plants in the true sense of the word) and the larger animals are called macrofauna.

Soil animals, especially, the earthworms and some insects tend to affect the soil favourably through their burrowing and feeding activities which tend to improve aeration and drainage through structural modifications of the soil solum. In general, they affect soil chemical properties to a lesser extent though their actions indirectly enhance microbial activities due to creation of a more favourable soil environment.

Soil Microorganisms

Soil microorganisms occur in huge numbers and display an enormous diversity of forms and functions. Major microbial groups in soil are bacteria (including actinomycetes), fungi, algae (including cyanobacteria) and protozoa. Because of their extremely small cell size (one to several micrometers), enormous numbers of soil microbes can occupy a relatively small volume, hence space is rarely a constraint on soil microbes. Soil microbes can occur in numbers ranging up to several million or more in a gram of fertile soil (a volume approximately that of a red kidney bean). Note that the bacteria are clearly the most numerous of the soil microbes. Perhaps more important than the numbers of the various soils microbes is the microbial biomass contributed by the respective groups. It is the soil fungi which tend to contribute the most biomass among the microbial groups. In fact, it is because of their large contribution to the biomass that they are generally regarded as being the dominant decomposer microbes in the soil. You might find it surprising that there are literally "tons" of microbes beneath your feet as you walk across a grassland in Africa or Australia or through a cornfield in the American Midwest. Interesting-ly, a fungus discovered in the state of Michigan may be one of the largest living organ-isms on the planet.

A fungus, *Armillaria bulbosa,* discovered in the U.S., in the state of Michigan, could turn out to be earth's largest creature or at least among the largest. Scientists discovered the fungus growing among the roots of hardwood trees in a forest. The microscopic, branched filaments (called hyphae) of the fungus occupy a 14.8 ha (37-acre) area of land. Careful genetic analysis has shown the filaments constitute a single organism. Fungi generally radiate outward in a circular pattern as they grow through the soil. In fact, the fairy rings of mushrooms (named because ancient peoples thought they represented the paths of fairies dancing in the night) often seen in lawns or on golf courses actually represent the outer boundary of a developing fungus. Scientists estimate that the portion of the Michigan fungus they have been able to identify may weigh as much as 100 tons, slightly less than a blue whale. Imagine the biochemical capacity of a soil microorganism this large!

The significance of these large amounts of microbial biomass in the soil lies not only in their large biochemical capacity, but in the phenomenal diversity of biochemical reactions attributed to the soil microbial population. It is worth remembering that soil microbes not only interact with other members of their own group, they also interact with other microbial groups. It is quite common to find, for example, that degradation of plant materials occurs much more quickly in the presence of the mixed soil population than it does when one or more groups of soil microbes have been eliminated from the system.

Soil life can be divided into trophic (*i.e.* feeding) levels. At the base of the trophic levels lies the soil microbial population which degrades plant, animal and microbial bodies, and also serves as the food source for some of the levels above it. For example, soil protozoa consume enormous numbers of bacteria and even some fungal spores. These in turn are consumed by still larger soil animals (nematodes, mites, etc.) which in turn are eaten by still larger animals (*e.g.* worms and insects). Thus, nutrients flow through this microbial food web which lies at the heart of controlling soil fertility and plant productivity in the absence of external inputs such as fertilizers. In fact, the role of soil microbes in degrading organic materials and thereby regenerating a supply of carbon dioxide for plants is perhaps their most vital global function.

Nutrient Cycling by Soil Microbes

Soil microbes exert much influence in controlling the quantities and forms of various chemical elements found in soil. Most notable are the cycles for carbon, nitrogen, sulfur and phosphorus, all of which are elements important in soil fertility, and as we know today, may be involved in global environmental phenomena. The mineralisation (*i.e.*, the conversion of organic forms of the elements to their inorganic forms) of organic materials by soil microbes liberates carbon dioxide, ammonium (which is rapidly converted to nitrate by soil microbes), sulfate, phosphate and inorganic forms of other elements. This is the basis of nutrient cycling in all major ecosystems of the world. John Burroughs once said, *"Without death and decay, how could life go on?"* No doubt, he was referring to the mineralisation of nutrients from dead animals and plants. We now know that soil microbes accomplish this task with remarkable zeal and that in the process a substantial part (perhaps as much as one third) of the decomposing materials are converted to the bodies of soil microbes. This pool of microbial biomass constitutes a portion of the soil organic matter which turns over (cycles) fairly quickly and therefore represents a "fertility buffer" in the soil. Don't forget that the liberation of carbon dioxide through microbial respiration makes possible the continued photosynthesis (*i.e.* carbon dioxide fixation) by algae and green plants which in turn produce more organic materials which may ultimately reach the soil, thereby completing the cycle.

In the world's agricultural soils, the source of our food supply, mineralisation of nitrogen by soil microbes is a most important process. In those soils not receiving external inputs of fertilizer nitrogen (*e.g.* most forested lands and many grasslands) the liberation of ammonium from organic debris makes possible the continued growth of new plant matter. Therefore, it is the soil microbial population which controls the productivity of these soils if other environmental factors (moisture, temperature) are suitable. In fact, fertilization of a soil represents our attempt to balance the competition between plants and soil microbes for available soil nitrogen. Nitrogen tied-up (assimilated into cell constituents) in microbial cells is not available for plants or other microbes until that tissue has been decomposed by other microbes. In other words, nitrogen contained in tissues is said to be immobilised. Microbes are the keys for the remobilisation of these nutrients. These mineralisation/ immobilisation phenomena are common to all the elements but typically they are only agriculturally important for the macronutrients such as nitrogen, phosphorus and sulfur. Aside from their role in controlling the rates of production of inorganic forms of nitrogen and sulfur, soil microbes, in particular soil bacteria, can control the forms of the ions in which these nutrients occur. For example, ammonium (NH_4^+) in the soil is usually rapidly oxidised by bacteria first to nitrite (NO_2^-) and then to nitrate (NO_3^-) which may readily leach through soil. Ammonium is oxidised to nitrite and then to nitrate by the bacteria *Nitrosomonas* and *Nitrobacter*, respectively. Thus, bacteria can influence the form and, thereby, the retention of nitrogen in the soil. Similarly, reduced sulfur compounds such as thiosulfate, elemental sulfur and even iron pyrite (FeS_2, "Fool's Gold") can be oxidised to sulfuric acid by soil bacteria. The bacteria which accomplish the oxidation of reduced nitrogen and sulfur compounds use these materials as energy sources to drive their metabolism. Unlike the decomposer microbes which use organic carbon compounds from organic matter for energy and to make cell matter (*e.g.* they are called heterotrophs), these specialised bacteria called chemoautotrophs obtain their carbon for cell synthesis from carbon dioxide or from dissolved carbonate.

There are many genera of bacteria that can oxidise reduced sulfur compounds. However, much of this activity, especially the oxidation of sulfur and pyrite, can be attributed to bacteria of the genus *Thiobacillus* (thio = sulfur; bacillus = rod-shaped bacterium). *Thiobacillus thiooxidans* can oxidise elemental sulfur to sulfuric acid. Sulfur, therefore, can be used to decrease the pH of an alkaline soil. *Thiobacillus ferrooxidans* attacks both the iron and sulfur in iron pyrite, generating sulfuric acid and dissolved iron in the process. This is also the basis of acid mine drainage associated with the mining of coal throughout the world.

The long-term application of ammonium-based fertilizers can likewise result in the acidification of agricultural soils through bacterial nitrification (the conversion of ammonium to nitrate with the concurrent production of acidity). Thus, we see that certain environmental problems can arise from the activities of these chemoautotrophic soil bacteria.

Another important aspect of nutrient cycling is that under certain circumstances nitrogen and sulfur may be converted to gaseous forms (volatilised) and lost to the atmosphere. Nitrogen in the form of nitrate can be converted to gases such as nitrous oxide (N_2O) and dinitrogen (N_2) through the process of denitrification (the bacterial reduction of NO_3^- to N_2O or N_2) by soil bacteria under anaerobic conditions. A consequence of denitrification is that nitrogen, a precious nutrient for plants, is lost from the soil. On the other hand, this process is a useful way to remove excess nitrate from wastewater.

Sulfur in the form of sulfate (SO_4^{-2}) is used by anaerobic bacteria like the genus *Desulfovibrio* which convert it to hydrogen sulfide gas (H_2S). Hydrogen sulfide reacts with metal ions and forms very insoluble metallic sulfides like pyrite (Fe_2S). In fact, it is probable that the pyrites associated with coal seams were deposited by the action of these bacteria eons ago. The black colour of salt marsh soils and the rotten egg smell associated with them are a result of the activities of the sulfate-reducing bacteria in these habitats. They attest to the occurrence of anaerobic conditions. Sulfur volatilisation from soil represents loss of a plant nutrient as well as a contribution of atmospheric sulfur which may contribute to the phenomenon of acid precipitation.

We mentioned above that nitrogen can be lost from agricultural soils as well as from other ecosystems. Fortunately, this "leak" in the terrestrial nitrogen cycle can be at least partially replaced through another important biological process called biological nitrogen fixation. In this process, which is unique to bacteria and a few other mi-crobes, notably the cyanobacteria (blue-green algae), atmospheric dinitrogen (N_2) is captured and converted to plant-available forms. Biological nitrogen fixation is carried out by free-living bacteria and cyanobacteria and by symbiotic microorganisms in a wide variety of mutualistically symbiotic associations with higher plants.

The most useful and probably the most widely recognised example of symbiotic nitrogen fixation is that of the *Rhizobium* - legume root-nodule symbiosis. Soil bacteria belonging to the genera *Rhizobium* and *Bradyrhizobium* (and a few others) are capable of inducing the formation of nodules on roots of specific legumes (plants like peas, beans, peanuts, soybeans, alfalfa etc.) and fixing large quantities of nitrogen in these structures. In the nodule, the bacteria are supplied with carbon sources (photosynthate from the plant) that they need in order to fix nitrogen. In return for this carbon, the bacteria fix atmospheric nitrogen which is converted to amino acids used by the plant for growth. The result of this unique plant-microbe partnership is that many legumes are self-sufficient for nitrogen, that is, they are nearly independent of a supply of nitrogen from the soil. It is no wonder that these plants are cultivated all over the world as sources of food, fibre and forage. Nearly two-thirds of the world's nitrogen supply is from biological nitrogen fixation. Legumes have been used since the beginning of recorded history as "soil improving" crops known as "green manures". Green manuring is the practice of growing a legume species for the sole purpose of returning it to the soil to serve as a source of nitrogen for an ensuing crop.

Soil Microbes and Bioremediation

We have touched on the remarkable metabolic diversity and capacity of the soil microflora. This capacity is increasingly being harnessed and put to good use by humans. A most beneficial spin-off from our understanding of the metabolism of soil microbes has been the development of methods for the bioremediation of soils contaminated by hazardous wastes or spilled petroleum products both on land and sea. Bioremediation may be defined as the controlled use of microorganisms for the destruction of chemical pollutants. A large number of processes have been developed to handle various wastes and for the cleanup of spilled organic materials. At the heart of all of these processes lies the premise that the metabolic activities of bacteria or fungi can be used to degrade many of the organic chemicals of commerce (solvents, pesticides, hydrocarbon fuels, etc.).

Either of two forms of bioremediation is commonly employed. In biostimulation the environment into which the material has been spilled or otherwise introduced is made favourable for the rapid development of microbes. Typically, this process involves adding sufficient nitrogen and phosphorus fertilizer to overcome nutrient limitations to microbial growth and providing some mechanism for increased aeration of the system. These practices encourage development of the indigenous microbial population which usually contains microbes able to degrade the compounds of interest. In the practice of bioaugmentation, an external microbial population is added in order to speed up the degradation process. Numerous microbes have been developed for such purposes. However, the full measure of the usefulness of such microbial products is not yet known. Some inoculants have reportedly enhanced the remediation process and others have had little or no effect on the process. It is probable that in due time useful microbial products or processes will be developed for use in the clean-up of oil or other chemical spills. What is certain is that successful bioremediation will require detailed knowledge of the factors which make some microbes more competitive than others in a given environment. Only when these details are established will we know how to use sound ecological principles to add microbes to these complex environments to insure their establishment and function in the clean-up process.

In March 1989, the Exxon Valdez oil tanker hit a reef in Prince William Sound, Alaska (USA) and released over 40 million litres of crude oil into the Sound within a 5-hour period. Over 1500 km (932 miles) of shoreline in the Sound and the Gulf of Alaska were contaminated to varying degrees by crude oil. The Exxon Valdez oil spill was a historic event because of the magnitude of the spill, the vastness and isolation of the area to be treated, and the large number of personnel and vehicles ultimately involved. The success of bioremediation, particularly in a climate as cold as Alaska's, prompted regulatory agencies in the United States to view bioremediation much more favourably over previous strategies of physical or chemical "entombment" (storage in cement tombs).

Because oil is inherently high in carbon and low in nitrogen and phosphorus, a portion of the shoreline was selected for biostimulation. After several potential fertilizer candidates were evaluated, a microemulsion, Inipol EAP22™ (henceforth, Inipol), was selected. Inipol (an "oleophilic" fertilizer) is a stable water-in-oil formulation that yields an N-P-K ratio of 7.3:0.8:0. The nitrogen source is urea and the phosphorus source is trilaureth (4)-phosphate.

At room temperature, Inipol has the consistency and appearance of honey, and it must be heated to 90 °C (194 °F) before it can be sprayed on the soil. Inipol was applied as a thin coat to the shore at a rate of 306 ml m^{-2} (0.27 quart per square yard). As the microemulsion mixed with the weathered crude oil, the crude oil destabilised Inipol to release its urea-N. In addition, a surface-active organic material (oleic acid) in Inipol served as a readily degradable carbon and energy source to increase the activity and number of indigenous hydrocarbon-degrading bacteria. When the oleic acid was depleted, the increased biomass of hydrocarbon-degrading bacteria supported enhanced biodegradation of the petroleum. Visual observations and chemical assays showed dramatic evidence that biostimulation contributed to the remediation of the site. Although passive bioremediation also undoubtedly occurred in the absence of the fertilizer nitrogen and phosphorus, the accelerated rate of biodegradation observed with Inipol was critical to a successful bioremediation effort.

SOIL FUNGI

Fungi are primarily organisms that cannot synthesise their own food and are dependent on complex organic substances for carbon. Specialized fungi can be pathogenic on the tissues of plants, while others form mutually beneficial relationships with plants and assist in direct nutrient supply to the plants (*e.g.*, mycorrhizal associations).

Many fungi play a very important role in the recycling of important chemical elements that would otherwise remain locked up in dead plants and animals. In the decomposition of plant debris, certain fungi are particularly important because of their ability to derive their carbon and energy requirements from the break down of dead and decaying plant cell walls, cellulose and lignin. They are much less dependent on water than other microorganisms, but interactions with other microbes, temperature and nutrient availability will have an effect on their activity. Fungal activity is greatest in decomposing leaves and wood, and tends to diminish in the later stages of decomposition when bacteria become more dominant.

Mycorrhizal Fungi

Mycorrhizas are associations between fungi and plant roots that can be beneficial to both the plant and the fungi. The fungi link the plant with soil by acting as agents of nutrient exchange. The fungi receive carbohydrates as energy from the host plant root whilst nutrients such as phosphorus and zinc are passed back into the plant roots from the soil. Mycorrhizal associations may also reduce attack from root pathogens and increase the tolerance of the plant to adverse conditions such as heavy metals, drought, and salinity. In general, mycorrhizas play an important role in sustainable plant productivity and maintenance of soil structure.

Mycorrhizal associations occur on almost all terrestrial plants and are not as plant-specific as other plant-microbe associations that formed between some plants (*e.g.,,* legumes) and bacteria (*e.g.,* rhizobia).

Different Types of Mycorrhizal Associations

There are four main kinds of mycorrhizal fungi: arbuscular, ectomycorrhizal, ericoid and orchid mycorrhiza. Most agricultural plants, vegetables and orchard plants form arbuscular mycorrhizal associations. Ectomycorrhizal associations are less common in disturbed ecosystems and are more common on perennial plants than annuals. Ornamental plants form associations from each of the four groups and orchid mycorrhizal associations are formed only by orchids.

Vesicular Arbuscular Micorrhizae

The most common type of mycorrhizas are the arbuscular mycorrhizas. They are found in natural ecosystems as well as in agricultural areas, are common on both perennials and annuals, and form associations with most agricultural plants. Two exceptions among agricultural crops are canola (oilseed rape) and lupin.

Arbuscules are believed to be the major site where the carbon and nutrient exchange between plant and fungus occurs. Named because of their "tree-like" structure, arbuscules are created by repeated branching of hyphae once they enter a cell within a plant root.

Vesicles are structures formed inside a cell within the plant's roots. They can be regular or irregular in shape and many times wider than the hyphae on which they form. Therefore, vesicles are usually very distinctive and in some species can resemble spores. These structures are known as a place for storing nutrients. Only three of the five genera of arbuscular mycorrhizas (Glomus, Acaulospora and Entrophosphora) form vesicles.

The fungi are grouped according to the size, shape and wall structure of their spores. The spores are approximately spherical or ovoid in shape and usually have thick walls which allow them to survive harsh environmental conditions. When soil conditions are favourable, the spores germinate and hyphae grow from the spore, entering roots and establishing mycorrhizal associations. Spores are between 30 and 500 micrometres in size - this is between 30 thousandths of a millimetre and half a millimetre. Identifying and classifying the spores is therefore something that must be done with the aid of a microscope.

Benefits of Mycorrhizal Associations

Mycorrhizal fungi are characterised by very thin hyphae, which are between 1 and 10 thousandths of a millimetre in width. These hyphae explore the soil for nutrients, transport them back to the host-plant, and help bind soil particles into aggregates. The hyphae form networks between neighbouring soil particles, between roots and soil particles, between roots on the same plant, and between roots of different plants (even different types of plants). They also form networks inside the roots they colonize. These networks of hyphae are also referred to as mycelium.

Mycorrhizas extend the volume of soil explored by the plant, a characteristic that is especially important for phosphorous which does not move in the soil solution as nitrogen does. There is some evidence that the fungi may help the plants tolerate drought. When phosphorus is scarce in soil, plants that have developed mycorrhizas on their root systems have greater access to and take up more phosphorus others. Trace elements, copper and zinc behave in a similar way to phosphorus in soil and plant roots must explore the soil to intercept them.

Although arbuscular mycorrhizas can enhance plant growth in phosphorous deficient soils, the extent to which this occurs in the field in agricultural and natural environments is difficult to measure.

Plant Pathogenic Fungi

Disease-causing microorganisms have always been inherent members of any living community. In natural ecosystems, characterised by uncontrolled and changeable conditions, their population growth is impeded by the scattered distribution of host plants and, in the case of fungal pathogens, by their dependence on rainfall at the time of spore germination. In managed systems, however, such as agriculture and horticulture, monocultures of crop plant species provide an unconstrained food supply for a pathogen. Irrigated systems also provide a constant supply of water which can enable spores to germinate and cause disease in accessible host plants.

Fungal Soil Pathogens

Climatic patterns can affect the types of fungal pathogens that are dominant in a region. For example, low fertility soils favour necrotrophic pathogens over biotrophic pathogens. Necrotrophic pathogens are distinguished from biotrophs because they kill host tissue prior to colonisation. Biotrophic pathogens include powdery mildew, downy mildew, rust, nematodes and viruses. Biotrophs live on living tissue and die when the host plant dies.

Soil Conditions Favour the Growth of Fungal Pathogens

The soil conditions that exist at the opening of the cropping season (warm-moist soils and low microbial activity) can favour the growth of a pathogen.

Reduced tillage practices help maintain infested residues at the surface of soil, increasing the damage to young seedlings. Conventional cultivations bury this inoculum source which gets broken-down more rapidly by soil microorganisms than when on the soil surface. Rotations with susceptible hosts can increase the inoculum potential of the pathogen in soil. Certain herbicides also increase the disease severity (e.g: the disease caused by the Take-all fungus and root rot caused by Rhizoctonia).

SOIL BACTERIA

Most bacteria in soil are about one micron in length or diameter (there are a thousand microns in a millimetre). Some are slightly larger than this, up to several microns, and in rare cases even larger. Their size varies with their environment. Bacteria in environments that have high levels of nutrients may be larger than those in nutrient poor conditions. The majority of bacteria in soil usually occur as single cells. Bacteria sometimes join together in chains or clusters. They mainly have one of two shapes - spheres (called cocci) and rods (called bacilli). Other bacteria have more varied shapes including spirals and long thin hyphae (although these are less common).

Importance of Bacteria in the Soil

Bacteria are able to perform an extremely wide range of chemical transformations, including degradation of organic matter, disease suppression, disease, and nutrient transformations inside roots (*e.g.,,* reducing bacteria in roots, bacteria cause nitrogen fixation).

Azobacter, for example, is a genus of free-living bacteria that converts atmospheric nitrogen into ammonium, making it available for plant use. This process may only take place, however, if the following conditions are met:

- An easily degradable carbon source is available
- Any nitrogen compounds such as ammonium or nitrate, are not already in present in substantial concentrations
- Soil pH levels are between 6 and 9
- High levels of phosphorus are present
- Very low levels of oxygen are present

Azobacter is inhibited by a large range of toxic mineral and organic compounds, but may tolerate relatively high salinity and their activities are enhanced in the presence of clays. In general, bacteria are the organisms in soil that are mainly responsible for transforming inorganic constituents from one chemical form to another. Their system of external digestion means that some of the metabolites released by the use of extracellular enzymes may be used by other organisms, such as plants. The bacteria gain nutrients and energy from these processes and provide other organisms with suitable forms of chemicals they require for their own processes. For example, in the conversions of nitrate to nitrite, sulphate to sulphide and ammonium to nitrite.

Where do Bacteria Live in Soil

Bacteria are aquatic organisms that live in the water-filled pore spaces within and between soil aggregates. As such, their activities are directly dependent on relatively high soil water contents. Bacteria are normally found on the surfaces of mineral or organic particles or congregate around particles of decaying plant and animal debris. Most are unable to move and hence, their dispersion is dependent on water movement, root growth or the activity of soil and other organisms.

RHIZOBIA

Rhizobia are one of the groups of microorganisms living in soil. They are single celled bacteria, approximately one thousandth of a millimetre in length. Rhizobia belong to a family of bacteria called Rhizobiaceae. There are a number of groups (genera and species) of bacteria in this family.

Rhizobia belong to a specific group of bacteria that form a mutually beneficial association, or symbiosis, with legume plants. These bacteria take nitrogen from the air (which plants cannot use) and convert it into a form of nitrogen called ammonium (NH_4^+), which plants can use. The nitrogenase enzyme controls the process, called nitrogen fixation, and these bacteria are often called "nitrogen fixers".

Rhizobia are found in soils of many natural ecosystems. They may also be present in agricultural areas where they are associated with both crop legumes (like soybean) and pasture legumes (like clover). Usually, the rhizobia in agricultural areas have been introduced at sowing by applying an inoculum to the exterior of the seeds as liquid formations or pellets.

Nodules Formed on the Roots of Legumes

The nodulation process is a series of events in which rhizobia interact with the roots of legume plants to form a specialised structure called a root nodule. These are visible, ball-like structures that are formed by the plant in response to the presence of the bacteria.

The process involves complicated signals between the bacteria and the host roots. In the first stages, the bacteria multiply near the root and then adhere to it. The small hairs on the root's surface curl around the bacteria and they enter the root. Alternatively, the bacteria may enter directly through points on the root surface. The method of entry of the bacteria into the root depends on the type of plant. Once inside the root, the bacteria multiply within thin threads. Signals stimulate cell multiplication of both the plant's cells and the bacteria and this repeated division results in a mass of root cells containing many bacterial cells. Some of these bacteria then change into a form that is able to convert gaseous nitrogen into ammonium nitrogen (that is, they can "fix" nitrogen). These bacteria are then called bacteroids.

The shape of the nodules is controlled by the plant and nodules can vary considerably - both in size and shape. Most plants need specific kinds of rhizobia to form nodules. The rhizobia that form nodules on peas, for example, cannot form nodules on clover. Nodulation can be impeded by low pH, Al toxicity, nutrient deficiencies, salinity, waterlogging, and the presence of root parasites such as nematodes or genetic incompatibility with the plant host.

Why are Nodules Pink Inside

The nitrogenase enzyme is extremely sensitive to oxygen and is only active at low oxygen levels or anaerobic conditions. The physical structure of the nodule acts as a barrier to oxygen and the enzyme leghaemaglobin binds oxygen and transports it away from nitrogenase to respiratory sites. Leghaemoglobin gives the inside of nodules their reddish-pink colouring.

Other Plant-bacterial Associations that Fix Atmospheric Nitrogen

Associations between bacteria and plants that fix atmospheric nitrogen include an association between species of Frankia bacteria and several tree species such as those of the genera Casuarina and Allocasuarina. Another example is between that of Azospirillum and grasses.

5

Microorganisms of the Soil

Raw organic matter in the soil is not directly used by the plants as food. It must be broken down first into humus and then into simpler products before it can be utilized. This work is done by different kinds of microorganisms in the soil.

The decomposition of organic matter forms part of the feeding and growth process of these microscopic plants and animals. Sugars, starches and proteins are broken down first; then cellulose and fatty substances (lipoids); and lastly lignin or woody substances. When the organic matter becomes chemically and biologically inert, the soil becomes infertile. However not all the soil organisms are beneficial.

There are certain bacteria which under anaerobic conditions of waterlogged soil cause denitrification releasing free nitrogen which gets lost into the air. Some others cause plant diseases. Intermediate products of bacterial decomposition of soil organic matter or certain organic products synthesized in the soil by some micro-organisms have important growth-promoting or plant protecting (*i.e.*, antibiotic) properties.

Pesticides are the chemical substances that kill pests and herbicides are the chemicals that kill weeds. In the context of soil, pests are fungi, bacteria insects, worms, and nematodes etc. that cause damage to field crops. Thus, in broad sense pesticides are insecticides, fungicides, bactericides, herbicides and nematicides that are used to control or inhibit plant diseases and insect pests. Although wide-scale application of pesticides and herbicides is an essential part of augmenting crop yields; excessive use of these chemicals leads to the microbial imbalance, environmental pollution and health hazards. An ideal pesticide should have the ability to destroy target pest quickly and should be able to degrade non-toxic substances as quickly as possible.

The ultimate "sink" of the pesticides applied in agriculture and public health care is soil. Soil being the storehouse of multitudes of microbes, in quantity and quality, receives the chemicals in various forms and acts as a scavenger of harmful substances. The efficiency and the competence to handle the chemicals vary with the soil and its physical, chemical and biological characteristics.

EFFECTS OF PESTICIDES

Pesticides reaching the soil in significant quantities have direct effect on soil microbiological aspects, which in turn influence plant growth.

Some of the most important effects caused by pesticides are:

- Alterations hi ecological balance of the soil microflora,
- Continued application of large quantities of pesticides may cause ever lasting changes in the soil microflora,
- Adverse effect on soil fertility and crop productivity,
- Inhibition of N2 fixing soil microorganisms such as *Rhizobium, Azotobacter, Azospirillum* etc. and cellulolytic and phosphate solubilizing microorg-anisms,
- Suppression of nitrifying bacteria, *Nitrosomonas* and *Nitrobacter* by soil fumigants ethylene bromide, Telone, and vapam have also been reported,
- Alterations in nitrogen balance of the soil,
- Interference with ammonification in soil,
- Adverse effect on mycorrhizal symbioses in plants and nodulation in legumes, and
- Alterations in the rhizosphere microflora, both quantitatively and qualitatively.

PERSISTENCE OF PESTICIDES IN SOIL

How long an insecticide, fungicide, or herbicide persists in soil is of great importance in relation to pest management and environmental pollution. Persistence of pesticides in soil for longer period is undesirable because of the reasons: a)Accumulation of the chemicals in soil to highly toxic levels,b) May be assimilated by the plants and get accumulated in edible plant products,c)Accumulation in the edible portions of the root crops, d)To be get eroded with soil particles and may enter into the water streams, and finally leading to the soil, water and air pollutions. The effective persistence of pesticides in soil varies from a week to several years depending upon structure and properties of the constituents in the pesticide and availability of moisture in soil. For instance, the highly toxic phosphates do not persist for more than three months while chlorinated hydrocarbon insecticides (eg. DOT, aldrin, chlordane etc) are known to persist at least for 4–5 years and some times more than 15 years.

From the agricultural point of view, longer persistence of pesticides leading to accumulation of residues in soil may result into the increased absorption of such toxic chemicals by plants to the level at which the consumption of plant products may prove deleterious/hazardous to human beings as well as livestock's. There is a chronic problem of agricultural chemicals, having entered in food chain at highly inadmissible levels in India, Pakistan, Bangladesh and several other developing countries in the world. For example, intensive use of DDT to control insect pests and mercurial fungicides to control diseases in agriculture had been known to persist for longer period and thereby got accumulated in the food chain leading to food contamination and health hazards. Therefore, DDT and mercurial fungicides has been, banned to use in agriculture as well as in public health department.

BIODEGRADATION OF PESTICIDES IN SOIL

Pesticides reaching to the soil are acted upon by several physical, chemical, and biological forces. However, physical and chemical forces are acting upon/degrading the pesticides to some extent, microorganism's plays major role in the degradation of pesticides. Many soil microorganisms have the ability to act upon pesticides and convert them into simpler non-toxic compounds. This process of degradation of pesticides and conversion into non-toxic compounds by microorganisms is known as "biodegradation". Not all pesticides reaching to the soil are biodegradable and such chemicals that show complete resistance to biodegradation are called "recalcitrant".

The chemical reactions leading to biodegradation of pesticides fall into several broad categories which are discussed in brief in the following paragraphs:

- *Detoxification*: Conversion of the pesticide molecule to a non-toxic compound. Detoxification is not synonymous with degradation. Since a single chance in the side chain of a complex molecule may render the chemical non-toxic.
- *Degradation*: The breaking down/transformation of a complex substrate into simpler products leading finally to mineralization. Degradation is often considered to be synonymous with mineralization, *e.g.* Thirum (fungicide) is degraded by a strain of*Pseudomonas* and the degradation products are dimethlamine, proteins, sulpholipaids, etc.
- *Conjugation (complex formation or addition reaction)*: In which an organism make the substrate more complex or combines the pesticide with cell metabolites. Conjugation or the formation of addition product is accomplished by those organisms catalyzing the reaction of addition of an amino acid, organic acid or methyl crown to the substrate, for *e.g.*, in the microbial metabolism of sodium dimethly dithiocar-bamate, the organism combines the fungicide with an amino acid molecule normally present in the cell and thereby inactivate the pesticides/chemical.
- *Activation*: It is the conversion of non-toxic substrate into a toxic molecule, for eg. Herbicide, 4-butyric acid (2, 4-D B) and the insecticide Phorate are transformed and activated microbiologically in soil to give metabolites that are toxic to weeds and insects.
- *Changing the spectrum of toxicity*: Some fungicides/pesticides are designed to control one particular group of organisms/pests, but they are metabolized to yield products inhibitory to entirely dissimilar

groups of organisms, for *e.g.* the fungicide PCNB fungicide is converted in soil to chlorinated benzoic acids that kill plants.

Biodegradation of pesticides/herbicides is greatly influenced by the soil factors like moisture, temperature, PH and organic matter content, in addition to microbial population and pesticide solubility. Optimum temperature, moisture and organic matter in soil provide congenial environment for the break down or retention of any pesticide added in the soil. Most of the organic pesticides degrade within a short period (3-6 months) under tropical conditions. Metabolic activities of bacteria, fungi and actinomycetes have the significant role in the degradation of pesticides.

CRITERIA FOR BIOREMEDIATION/BIODEGRADATION

For successful biodegradation of pesticide in soil, following aspects must be taken into consideration:

- Organisms must have necessary catabolic activity required for degradation of contaminant at fast rate to bring down the concentration of contaminant,
- The target contaminant must be bioavailability,
- Soil conditions must be congenial for microbial/plant growth and enzymatic activity and
- Cost of bioremediation must be less than other technologies of removal of contaminants.

According to Gales (1952) principal of microbial infallibility, for every naturally occurring organic compound there is a microbe/enzyme system capable its degradation.

STRATEGIES FOR BIOREMEDIATION

For the successful biodegradation/bioremediation of a given contaminant following strategies are needed:

- *Passive/ intrinsic Bioremediation*: It is the natural bioremediation of contaminant by tile indigenous microorganisms and the rate of degradation is very slow.
- *Biostimulation*: Practice of addition of nitrogen and phosphorus to stimulate indigenous microorganisms in soil.
- *Bioventing*: Process/way of Biostimulation by which gases stimulants like oxygen and methane are added or forced into soil to stimulate microbial activity.
- *Bioaugmentation*: It is the inoculation/intro duction of microorganisms in the contamin ated site/soil to facilitate biodegradation.

ROLE AND APPLICATION

The concept of effective microorganisms (EM) was developed by Professor Teruo Higa, University of the Ryukyus, Okinawa, Japan (Higa, 1991; Higa and Wididana, 1991a). EM consists of mixed cultures of beneficial an naturally-occurring microorganisms that can be applied as inoculants to increase the microbial diversity of soils and plant. Research has shown that the inoculation of EM cultures to the soil/plant ecosystem can improve soil quality, soil health, and the growth, yield, and quality of crops.

EM contains selected species of microorganisms including predominant populations of lactic acid bacteria and yeasts and smaller numbers of photosynthetic bacteria, actinomycetes and other types of organisms. All of these are mutually compatible with one another and can coexist in liquid culture.

EM is not a substitute for other management practices. It is, however, an added dimension for optimizing our best soil and crop management practices such as crop rotations, use of organic amendments, conservation tillage, crop residue recycling, and biocontrol of pests. If used properly, EM can significantly enhance the beneficial effects of these practices.

Throughout the discussion which follows, we will use the term "beneficial microorganisms" In a general way to designate a large group of often unknown or ill-defined microorganisms that interact favourably in soils and with plants to render beneficial effects which are sometimes difficult to predict. We use the term "effective microorganisms" or EM to denote specific mixed cultures of known, beneficial microorganisms that are being used effectively as microbial inoculants.

Conceptual design is important in developing new technologies for utilizing beneficial and effective microorganisms for a more sustainable agriculture and environment. The basis of a conceptual design is imply to first conceive an ideal or model and then to devise a strategy and method for achieving the reality. However it is necessary to carefully coordinate the materials, the environment, and the technologies constituting the method. Moreover one should adopt a philosophical attitude in applying microbial technologies to agricultural production and conservation systems.

There are many opinions on what an ideal agricultural system is. Many would agree that such an idealized system should produce food on a long-term sustainable basis. Many would also insist that it should maintain and improve human health, be economically and spiritually beneficial to both producers and consumers, actively preserve and protect the environment, be self-contained and regenerative, and produce enough food for an increasing world population.

UTILIZATION AND RECYCLING OF ENERGY

Agricultural production begins with the process of photosynthesis by green plants which requires solar energy, water, and carbon dioxide. It occurs through the plants ability to utilize solar energy in "fixing" atmospheric carbon into carbohydrates. The energy obtained is used for further biosynthesis in the plant, including essential amino acids and proteins. The materials used for agricultural production are abundantly available with little initial cost. However, when it is observed as an economic activity, the fixation of carbon by photosynthesis has an extremely low efficiency mainly because of the low utilization rate of solar energy by green plants. Therefore, an integrated approach is needed to increase the level of solar energy utilization by plants so that greater amounts of atmospheric carbon can be converted into useful substrates.

Although the potential utilization rate of solar energy by plants has been estimated theoretically at between 10 and 20%, the actual utilization rate is less than 1%. Even the utilization rate of C4 plants, such as sugar cane whose photosynthetic efficiency is very high, barely exceeds 6 or 7% during the maximum growth period. The utilization rate is normally less than 3% even for optimum crop yields.

Past studies have shown that photosynthetic efficiency of the chloroplasts of host crop plants cannot be increased much further; this means that their biomass production has reached a maximum level. Therefore, the best opportunity for increasing biomass production is to somehow utilize the visible light, which chloroplasts cannot presently use, and the infrared radiation; together, these comprise about 80% of the total solar energy. Also, we must explore ways of recycling organic energy contained in plant and animal residues through direct utilization of organic molecules by plants.

Thus, it is difficult to exceed the existing limits of crop production unless the efficiency of utilizing solar energy is increased, and the energy contained in existing organic molecules (amino acids, peptides and carboh ydrates) is utilized either directly or indirectly by the plant. This approach could help to solve the problems of environmental pollution and degradation caused by the misuse and excessive application of chemical fertilizers and pesticides to soils. Therefore, new technologies that can enhance the economic-viability of farming systems with little or no use of chemical fertilizers and pesticides are urgently needed and should be a high priority of agricultural research both now and in the immediate future.

PRESERVATION OF NATURAL RESOURCES AND THE ENVIRONMENT

The excessive erosion of topsoil from farmland caused by intensive tillage and row-crop production has caused extensive soil degradation and also contributed to the pollution of both surface and groundwater. Organic wastes from animal production, agricultural and marine processing industries, and municipal wastes (*e.i.*, sewage and garbage), have become major sources of environmental pollution in both developed and developing countries. Furthermore, the production of methane from paddy fields and ruminant animals and of carbon dioxide from the burning of fossil fuels, land clearing and organic matter decomposition have been linked to global warming as "greenhouse gases". Chemical-based, conventional systems of agricultural production have created many sources of pollution that, either directly or indirectly, can contribute to degradation of the environment and destruction of our natural resource base. This situation would change significantly if these pollutants could be utilized in agricultural production as sources of energy.

Therefore, it is necessary that future agricultural technologies be compatible with the global ecosystem and with solutions to such problems in areas different from those of conventional agricultural technologies. An area that appears to hold the greatest promise for technological advances in crop production, crop protection, and natural resource conservation is that of beneficial and effective microorganisms applied as soil, plant and environmental inoculants.

BENEFICIAL AND EFFECTIVE MICROORGANISMS FOR A SUSTAINABLE AGRICULTURE

Agriculture in a broad sense, is not an enterprise which leaves everything to nature without intervention. Rather it is a human activity in which the farmer attempts to integrate certain agroecological factors and production inputs for optimum crop and livestock production. Thus, it is reasonable to assume that farmers should be interested in ways and means of controlling beneficial soil microorganisms as an important component of the agricultural environment. Nevertheless, this idea has often been rejected by naturalists and proponents of nature farming and organic agriculture. They argue that beneficial soil microorganisms will increase naturally when organic amendments are applied to soils as carbon, energy and nutrient sources. This indeed may be true where an abundance of organic materials are readily available for recycling which often occurs in small-scale farming. However, in most cases, soil microorganisms, beneficial or harmful, have often been controlled advantageously when crops in various agroecological zones are grown and cultivated in proper sequence (*i.e.*, crop rotations) and without the use of pesticides. This would explain why scientists have long been interested in the use of beneficial microorganisms as soil and plant inoculants to shift the microbiological equilibrium in a way that enhances soil quality and the yield and quality of crops.

Most would agree that a basic rule of agriculture is to ensure that specific crops are grown according to their agroclimatic and agroecological requirements. However, in many cases the agricultural economy is based on market forces that demand a stable supply of food, and thus, it becomes necessary to use farmland to its full productive potential throughout the year.

The purpose of crop breeding is to improve crop production, crop protection, and crop quality. Improved crop cultivars along with improved cultural and management practices have made it possible to grow a wide variety of agricultural and horticultural crops in areas where it once would not have been culturally or economically feasible. The cultivation of these crops in such diverse environments has contributed significantly to a stable food supply in many countries. However, it is somewhat ironic that new crop cultures are almost never selected with consideration of their nutritional quality or bioavailability after ingestion.

As will be discussed later, crop growth and devel opment are closely related to the nature of the soil microflora, especially those in close proximity to plant roots, *i.e.*, the rhizosphere. Thus, it will be difficult to overcome the limitations of conventional agricultural technologies without controlling soil microorganisms. This particular tenet is further reinforced because the evolution of most forms of life on earth and their environments are sustained by microorganisms. Most biological activities are influenced by the state of these invisible, mi nuscule units of life. Therefore, to significantly increase food production, it is essential to develop crop cultivars with improved genetic capabilities (*i.e.*, greater yield potential, disease resistance, and nutritional quality) and with a higher level of environmental competitiveness, particularly under stress conditions (*i.e.*, low rainfall, high temperatures, nutrient deficiencies, and agressive weed growth).

To enhance the concept of controlling and utilizing beneficial microorganisms for crop production and protection, one must harmoniously integrate the essential components for plant growth and yield including light (intensity, photoperiodicity and quality), carbon dioxide, water, nutrients (organic-inorganic) soil type, and the soil microflora. Because of these vital interrelationships, it is possible to envision a new technology and a more energy-efficient system of biological production.

Low agricultural production efficiency is closely related to a poor coordination of energy conversion which, in turn, is influenced by crop physiological factors, the environment, and other biological factors including soil microorganisms. The soil and rhizosphere microflora can accelerate the growth of plants and enhance their resistance to disease and harmful insects by producing bioactive substances. These microorganisms maintain the growth environment of plants, and may have secondary effects on crop quality. A wide range of results are possible depending on their predominance and activities at any one time.

Nevertheless, there is a growing consensus that it is possible to attain maximum economic crop yields of high quality, at higher net returns, without the application of chemical fertilizers and pesticides. Until recently, this was not thought to be a very likely possibility using conventional agricultural methods.

However, it is important to recognize that the best soil and crop management practices to achieve a more sustainable agriculture will also enhance the growth, numbers and activities of beneficial soil microorganisms that, in turn, can improve the growth, yield and quality of crops.

KINDS OF SOIL ORGANISMS

The soil microorganisms have been classified as:

Microflora:

- Bacteria,
- Actinomycetes,
- Fungi, and
- Algae.

Microfauna:

- Protozoa, and
- Nematodes.

Besides microfauna the soil harbours a large number of worms and insects of different kinds and sizes.

Density of Population

The soil organisms vary in number from a few per hectare to many millions per gramme of soil. The density of population is determined by food supply, moisture, temperature, physical condition and the reaction of the soil. In neutral soils, bacteria dominate over other types of microscopic life. If the soil is acidic and rich in organic matter, fungi predominate. Algae abound on the soil in constantly moist or shady situations.

Under favourable conditions, the bacteria multiply enormously. In sandy desert soils and under waterlogged conditions they are very scarce.

As among the soil microflora, it is estimated that bacteria form about 90% of the total population, actinimycetes about 9%, and fungi and algae together about one per cent. Among the soil fauna, protozoa are the most abundant, followed by nematodes, worms, and insects.

Bacterial Activity

The soil microflora typically produce ammonia from organic compounds when they set free more of nitrogen than they can assimilate and convert into their own protoplasms. Ammonia so released is converted into nitrites by one group of organisms called Nitrosomonas and further converted into nitrates by another group of organisms called Nitrobacter. The process of conversion of nitrogen to nitrates is called nitrification. The nitrate-forming bacteria are generally confined to the top 25 to 30 cm of the soil, where the content of organic matter is also more.

These organisms are most active between 25 and 38 degrees Celsius and under favourable conditions of tillage, aeration, neutral soil reaction, and moisture content at field capacity. The bacteria cease or reduce their activity when the pH value of the soil falls below 5. They however become active again when the pH is raised about 6 by liming. The large supplies of carbon in the raw organic matter favour an enormous multiplication of bacteria which use up the nitrates released by nitrifying bacteria. A similar effect may also be produced by heavy green-manuring. This problem can be dealt with either by adding sufficient fertilizer nitrogen to the organic material to provide for the nitrogen needs of the micro-organisms and the main crop or by incorporating the organic matter into the soil sufficiently early to complete the decomposition process well before the sowing of the crop.

Two other groups of bacteria intimately linked with the nitrogen problem of the soil take up free nitrogen from the air and convert it into nitrogenous compounds for the use of crop plants. One group (genus rhizobium) functions in symbiosis with leguminous plants and the other genus aztobacter fixes free nitrogen independently of the legumes. The symbiotic bacteria penetrate the rootlets and induce the plants to produce nodules at the point of entry. The bacteria grow and multiply inside the nodules, getting their carbohydrate and mineral food from the plant and nitrogen from the air to nitrogenous organic compounds, *e.g.,* proteins, which then become available to the host plant. The amount of nitrogen added to the soil varies from 50-150 kg per hectare. In the absence of the nitrogen-fixing bacteria no nodules will be formed by the leguminous plants and no atmospheric nitrogen will be fixed. It has been found that the nodule organisms belong to several distinct strains, each of which is capable of producing nodules on a specific group of legumes. It has also been discovered that the presence of the most efficient bacteria for a given legume crop can be assured by inoculating the seed with a culture of the proper strain artificially grown in the laboratory.

Azotobacter and other non-symbiotic nitrogen-fixing bacteria work independently of any host crop. Under optimum laboratory conditions, Azotobacter has been found to fix a considerable amount of nitrogen.

Actinomycetes

They are similar in size to bacteria but resemble moulds in their growth and physiology. They can grow in the deeper layers of the soil and under drier conditions and need less nitrogen. They are also said to synthesize chemical substances which produce the familiar odour when the cultivated soil receives its first shower of rain.

Fungi

These organisms produce microscopic threads called mycelia which may be found in the disintegrating organic matter on the surface of the soil or on the plant roots in the upper strata below the surface. Many of the fungi are harmless saprophytes living on dead organic matter; others are parasites which attack live plants and produce highly destructive epidemic diseases. Still others are parasites that can also live in the absence of host plants.

Algae

They are microscopic or larger plants containing chlorophyll. They are found in large numbers in the top layer of constantly moist soils such as paddy fields. Some of them are capable of fixing atmospheric nitrogen.

Protozoa

Soil protozoa are unicellular animals feeding either on soil organic matter or on bacteria thus regulating the number of the latter in the soil. The larger forms are found in the very wet conditions or in the swamps.

FUNCTIONS OF MICROORGANISMS. PUTREFACTION, FERMENTATION, AND SYNTHESIS

Soil microorganisms can be classified into decomposer and synthetic microorganisms. The decomposer microorganisms are subdivided into groups that perform oxidative and fermentative decomposition. The fermentative group is further divided into useful fermentation (simply called fermentation) and harmful fermentation (called putrefaction). The synthetic microorganisms can be sub-divided into groups having the physiological abilities to fix atmospheric nitrogen into amino acids and/or carbon dioxide into simple organic molecules through photosynthesis.

Fermentation is an anaerobic process by which facultative microorganisms (*e.g.*, yeasts) transform complex organic molecules (*e.g.*, carbohydrates) into simple organic compounds that often can be absorbed directly by plants. Fermentation yields a relatively small amount of energy compared with aerobic decomposition of the same substrate by the same group of microorganisms. Aerobic decomposition results in complete oxidation of a substrate and the release of large amounts of energy, gas, and heat with carbon dioxide and water as the end products. Putrefaction is the process by which facultative heterotrophic microorganisms decompose proteins anaerobically, yielding malodorous incompletely oxidized, metabolites (*e.g.*, ammonia, mercaptans and indole) that are often toxic to plants and animals.

The term "synthesis" as used here refers to the biosynthetic capacity of certain microorganisms to derive metabolic energy by "fixing" atmospheric nitrogen and/or carbon dioxide. In this context we refer to these as "synthetic" microorganisms, and if they should become a predominant part of the soil microflora, then the soil would be termed a "synthetic" soil.

Nitrogen-fixing microorganisms are highly diverse, ranging from "free-living" autotrophic bacteria of the genus Azotobacter to symbiotic, heterotrophic bacteria of the genus Rhizobium, and blue-green algae (now mainly classified as blue-green bacteria), all of which function aerobically. Photosynthetic microorganisms fix atmospheric carbon dioxide in a manner similar to that of green plants. They are also highly diverse, ranging from blue-green algae to green algae that perform complete photosynthesis aerobically to photosynthetic bacteria which perform incomplete photosynthesis anaerobically.

RELATIONSHIPS BETWEEN PUTREFACTION, FERMENTATION, AND SYNTHESIS

The processes of putrefaction, fermentation, and synthesis proceed simultaneously according to the appropriate types and numbers of microorganisms that are present in the soil. The impact on soil quality attributes and related soil properties is determined by the dominant process. The production of organic substances by microorganisms results from the intake of positive ions, while decomposition serves to release these positive ions. Hydrogen ions play a pivotal role in these processes.

A problem occurs when hydrogen ions do not recombine with oxygen to form water but are utilized to produce methane, hydrogen sulfide, ammonia, mercaptans and other highly reduced putrefactive substances most of which are toxic to plants and produce malodors. If a soil is able to absorb the excess hydrogen ions during periods of soil anaerobiosis and if synthetic microorganisms such as photosynthetic bacteria are present, they will utilize these putrefactive substances and produce useful substrates from them which helps to maintain a healthy and productive soil.

The photosynthetic bacteria, which perform incomplete photosynthesis anaerobically, are highly desirable, beneficial soil microorganisms because they are able to detoxify soils by transforming reduced, putrefactive substances such as hydrogen sulfide into useful substrates. This helps to ensure efficient utilization of organic matter and to improve soil fertility. Photosynthesis involves the photo-catalyzed splitting of water which yields molecular oxygen as a by-product. Thus, these microorganisms help to provide a vital source of oxygen to plant roots. Reduced compounds such as methane and hydrogen sulfide are often produced when organic materials are decomposed under anaerobic conditions. These compo unds are toxic and can greatly suppress the activities of nitrogen-fixing microorganisms. However, if synthetic microorganisms, such as photosynthetic bacteria that utilize reduced substances, are present in the soil, oxygen deficiencies are not likely to occur. Thus, nitrogen-fixing microorganisms, coexisting in the soil with photosy nthetic bacteria, can function effectively in fixing atmospheric nitrogen even under anaerobic conditions.

Photosynthetic bacteria not only perform photos ynthesis but can also fix-nitrogen. Moreover, it has been shown that, when they coexist, in soil with species of Azotobacter, their ability to fix nitrogen is enhanced. This then is an example of a synthetic soil. It also suggests that by recognizing the role, function, and mutual compatibility of these two bacteria and utilizing them effectively to their full potential, soils can be induced to a greater synthetic capacity. Perhaps the most effective synthetic soil system results from the enhancement of zymogenic and synthetic microorganisms; this allows fermentation to become dominant over putrefaction and useful synthetic processes to proceed.

CLASSIFICATION OF SOILS BASED ON THE FUNCTIONS OF MICROORGANISMS

As discussed earlier, soils can be characterized according to their indigenous microflora which perform putrefactive, fermentative, synthetic and zymogenic reactions and processes. In most soils, these three functions are going on simultaneously with the rate and extent of each determined by the types and numbers of associated microorganisms that are actively involved at any one time.

Disease-Inducing Soils. In this type of soil, plant pathogenic microorganisms such as Fusarium fungi can comprise 5 to 20 per cent of the total microflora if fresh organic matter with a high nitrogen content is applied to such a soil, incompletely oxidized products can arise that are malodorous and toxic to growing plants. Such soils tend to cause frequent infestations of disease organisms, and harmful insects. Thus, the application of fresh organic matter to these soils is often harmful to crops. Probably more than 90 per cent of the agricultural land devoted to crop production worldwide can be classified as having disease-inducing soil. Such soils generally have poor physical properties, and large amounts of energy are lost as "greenhouse" gases, particularly in the case of rice fields. Plant nutrients are also subject to immobilization into unavailable forms.

Disease-Suppressive Soils

The microflora of disease-suppressive soils is usually dominated by antagonistic microorganisms that produce copious amounts of antibiotics. These include fungi of the genera Penicillium, Trichoderma, and Aspergillus, and actinomycetes of the genus Streptomyces. The antibiotics they produce can have biostatic and biocidal effects on soil-borne plant pathogens, including Fusarium which would have an incidence in these soils of less than 5 per cent. Crops planted in these soils are rarely affected by diseases or insect pests. Even if fresh organic matter with a high nitrogen content is applied, the production of putrescent substances is very low and the soil has a pleasant earthy odor after the organic matter is decomposed.

These soils generally have excellent physical properties; for example, they readily, form water-stable aggregates and they are well-aerated, and have a high permeability to both air and water. Crop yields in the disease-suppressive soils are often slightly lower than those in synthetic soils. Highly acceptable crop yields are obtained whenever a soil has a predominance of both disease-suppressive and synthetic microorganisms.

Zymogenic Soils

These soils are dominated by a microflora that can perform useful kinds of fermentations, *i.e.*, the breakdown of complex organic molecules into simple organic substances and inorganic materials. The organisms can be either obligate or facultative anaerobes. Such fermentation-producing microorganisms often comprise the microflora of various organic materials, *i.e.*, crop residues, animal manures, green manures and municipal wastes including composts. After these amendments are applied to the soil, their number: and fermentative activities can increase dramatically and overwhelm the indigenous soil microflora for an indefinite period.

While these microorganisms remain predominant, the soil can be classified as a zymogenic soil which is generally characterized by,

- Pleasant, fermentative odors especially after tillage,
- Favourable soil physical properties (*e.g.*, Increased aggregate stability, permeability, aeration and decreased resistance to tillage
- Large amounts of inorganic nutrients, amino acids, carbohydrates, vitamins and other bioactive substances which can directly or indirectly enhance the growth, yield and quality of crops,
- Low occupancy of Fusarium fungi which is usually less than 5 per cent, and
- Low production of greenhouse gases (*e.g.*, methane, ammonia, and carbon dioxide) from croplands, even where flooded rice is grown.

Synthetic Soils

These soils contain significant populations of microorganisms which are able to fix atmospheric nitrogen and carbon dioxide into complex molecules such as amino acids, proteins and carbohydrates. Such microorganisms include photosynthetic bacteria which perform incomplete photosynthesis anaerobically, certain Phycomycetes (fungi that resemble algae), and both green algae and blue—green algae which function aerobically. All of these are photosynthetic organisms that fix atmospheric nitrogen. If the water content of these soils is stable, their fertility can be largely maintained by regular additions of only small amounts of organic materials. These soils have a low Fusarium occupancy and they are often of the disease-suppressive type. The production of gases from fields where synthetic soils are present is minimal, even for flooded rice.

This is a somewhat simplistic classification of soils based on the functions of their predominant types of microorganisms, and whether they are potentially beneficial or harmful to the growth and yield of crops. While these different types of soils are described here in a rather idealized manner, the fact is that in nature they are not always clearly defined because they often tend to have some of the same characteristics. Nevertheless, research has shown that a disease-inducing soil can be transformed into disease-suppressing, zymogenic and synthetic soils by inoculating the problem soil with mixed cultures of effective microorganisms. Thus it is somewhat obvious that the most desirable agricultural soil for optimum growth, production, protection, and quality of crops would be the composite soil indicated in Fig., *i.e.*, a soil that is highly zymogenic and synthetic, and has an established disease-suppressive capacity. This then is the principle reason for seeking ways and means of controlling the microflora of agricultural soils.

CONTROLLING THE SOIL MICROFLORA

PRINCIPLES AND STRATEGIES

Principles of Natural Ecosystems and the Application of Beneficial and Effective Microorganisms. The misuse and excessive use of chemical fertilizers and pesticides have often adversely affected the environment and created many a) food safety and quality and b) human and animal health problems. Consequently, there has been a growing interest in nature farming and organic agriculture by consumers and environmentalists as possible alternatives to chemical-based, conventional agriculture. Agricultural systems which conform to the principles of natural ecosystems are now receiving a great deal of attention in both developed and developing countries. A number of books and journals have recently been published which deal with many aspects of natural farming systems. New concepts such as alternative agriculture, sustainable agriculture, soil quality, integrated pest management, integrated nutrient management and even beneficial microorganisms are being explored by the agricultural research establishment. Although these concepts and associated methodologies hold considerable promise, they also have limitations. For example, the main limitation in using microbial inoculants is the problem of reproducibility and lack of consistent results.

Unfortunately certain microbial cultures have been promoted by their suppliers as being effective for controlling a wide range of soil-borne plant diseases when in fact they were effective only on specific pathogens under very specific conditions. Some suppliers have suggested that their particular microbial inoculant is akin to a pesticide that would suppress the general soil microbial population while increasing the population of a specific beneficial microorganism. Nevertheless, most of the claims for these single-culture microbial inoculants are greatly exaggerated and have not proven to be effective under field conditions. One might speculate that if all of the microbial cultures and inoculants that are available as marketed products were used some degree of success might be achieved because of the increased diversity of the soil microflora and stability that is associated with mixed cultures. While this, of course, is a hypothetical example, the fact remains that there is a greater likelihood of controlling the soil microflora by introducing mixed, compatible cultures rather than single pure cultures.

Even so, the use of mixed cultures in this approach has been criticized because it is difficult to demonstrate conclusively which microorganisms are responsible for the observed effects, how the introduced microorganisms interact with the indigenous species, and how these new associations affect the soil/plant environment. Thus, the use of mixed cultures of beneficial microorganisms as soil inoculants to enhance the growth, health, yield, and quality of crops has not gained widespread acceptance by the agricultural research establishment because conclusive scientific proof is often lacking.

The use of mixed cultures of beneficial micr oorganisms as soil inoculants is based on the principles of natural ecosystems which are sustained by their constituents; that is, by the quality and quantity of their inhabitants and specific ecological parameters, *i.e.*, the greater the diversity and number of the inhabitants, the higher the order of their interaction and the more stable the ecosystem. The mixed culture approach is simply an effort to apply these principles to natural systems such as agricultural soils, and to shift the microbiological equilibrium in favour of increased plant growth, production and protection.

It is important to recognize that soils can vary tremendously as to their types and numbers of microor-ganisms. These can be both beneficial and harmful to plants and often the predominance of either one depends on the cultural and management practices that are applied. It should also be emphasized that most fertile and productive soils have a high content of organic matter and, generally, have large, populations of highly diverse microorganisms (*i.e.*, both species and genetic diversity). Such soils will also usually have a wide ratio of beneficial to harmful microorganisms.

CONTROLLING THE SOIL MICROFLORA FOR OPTIMUM CROP PRODUCTION AND PROTECTION

The idea of controlling and manipulating the soil microflora through the use of inoculants organic amendments and cultural and management practices to create a more favourable soil microbiological environment for optimum crop production and protection is not new. For almost a century, microbiologists have known that organic wastes and residues, including animal manures, crop residues, green manures, municipal wastes (both raw and composted), contain their own indigenous populations of microorganisms often with broad physiological capabilities.

It is also known that when such organic wastes and residues are applied to soils many of these introduced microorganisms can function as biocontrol agents by controlling or suppressing soil-borne plant pathogens through their competitive and antagonistic activities. While this has been the theoretical basis for controlling the soil microflora, in actual practice the results have been unpredictable and inconsistent, and the role of specific microorganisms has not been well-defined.

For, many years microbiologists have tried to culture beneficial microorganisms for use as soil inoculants to overcome the harmful effects of phytopathogenic organisms, including bacteria, fungi and nematodes. Such attempts have usually involved single applications of pure cultures of microorganisms which have been largely unsuccessful for several reasons. First, it is necessary to thoroughly understand the individual growth and survival characteristics of each particular beneficial micro organism, including their nutritional and environmental requirements. Second, we must understand their ecological relationships and interactions with other microorganisms, including their ability to coexist in mixed cultures and after application to soils.

There are other problems and constraints that have been major obstacles to controlling the microflora of agricultural soils. First and foremost is the large number of types of microorganisms that are present at any one time, their wide range of physiological capabilities, and the dramatic fluctuations in their populations that can result from man's cultural and management practices applied to a particular farming system. The diversity of the total soil microflora depends on the nature of the soil environment and those factors which affect the growth and activity of each individual organism including temperature, light, aeration, nutrients, organic matter, pH and water. While there are many microorganisms that respond positively to these factors, or a combination thereof, there are many that do not. Microbiologists have actually studied relatively few of the microorganisms that exist in most agricultural soil, mainly because we don't know how to culture them; *i.e.*, we know very little about their growth, nutritional, and ecological requirements.

The "diversity" and "population" factors associated with the soil microflora have discouraged scientists from conducting research to develop control strategies. Many believe that, even when beneficial microorganisms are cultured and inoculated into soils, their number is relatively small compared with the indigenous soil inhabitants, and they would likely be rapidly overwhelmed by the established soil microflora. Consequently, many would argue that even if the application of beneficial microorganisms is successful under limited conditions (*e.g.*, in the laboratory) it would be virtually impossible to achieve the same success under actual field conditions. Such thinking still exists today, and serves as a principle constraint to the concept of controlling the soil microflora.

It is noteworthy that most of the microorganisms encountered in any particular soil are harmless to plants with only a relatively few that function as plant pathogens or potential pathogens. Harmful microorganisms become dominant if conditions develop that are favourable to their growth, activity and reproduction. Under such conditions, soil-borne pathogens (*e.g.*, fungal pathogens) can rapidly increase their populations with devastating effects on the crop. If these conditions change, the pathogen population declines just as rapidly to its original state. Conventional farming systems that tend towards the consecutive planting of the same crop (*i.e.*, monoculture) necessitate the heavy use of chemical fertilizers and pesticides. This, in turn, generally increases the probability that harmful, disease-producing, plant pathogenic microorganisms will become more dominant in agricultural soils. Chemical-based conventional farming methods are not unlike symptomatic therapy. Examples of this are applying fertilizers when crops show symptoms of nutrient-deficiencies, and applying pesticides whenever crops are attacked by insects and diseases. In efforts to control the soil microflora some scientists feel that the introduction of beneficial microorganisms should follow a symptomatic approach. However, we do not agree. The actual soil conditions that prevail at any point in time may be most unfavourable to the growth and establishment of laboratory-cultured, beneficial microorganisms.

To facilitate their establishment, it may require that the farmer make certain changes in his cultural and management practices to induce conditions that will:

(a) Allow the growth and survival of the inoculated microorganisms and

(b) Suppress the growth and activity of the indigenous plant pathogenic microorganisms.

An example of the importance of controlling the soil microflora and how certain cultural and management practices can facilitate such control is useful here. Vegetable cultivars are often selected on their ability to grow and produce over a wide range of temperatures. Under cool, temperate conditions there are generally few pest and disease problems. However, with the onset of hot weather, there is a concomitant increase in the incidence of diseases and insects making it rather difficult to obtain acceptable yields without applying pesticides. With higher temperatures, the total soil microbial population increases as does certain plant pathogens such as Fusarium, which is one of the main putrefactive, fungal pathogens in soil. The incidence and destructive activity of this pathogen can be greatly minimized by adopting reduced tillage methods and by shading techniques to keep the soil cool during hot weather. Another approach is to inoculate the soil with beneficial, antagonistic, antibiotic-producing microorganisms such as actinomycetes and certain fungi.

APPLICATION OF BENEFICIAL AND EFFECTIVE MICROORGANISMS:A NEW DIMENSION

Many microbiologists believe that the total number of soil microorganisms can be increased by applying organic amendments to the soil. This is generally true because most soil microorganisms are heterotrophic, *i.e.*, they require complex organic molecules of carbon and nitrogen for metabolism and biosynthesis. Whether the regular addition of organic wastes and residues will greatly increase the number of beneficial soil microorganisms in a short period of time is questionable. However, we do know that heavy applications of organic materials, such as seaweed, fish meal, and chitin from crushed crab shells, not only helps to balance the micronutrient content of a soil but also increases the population of beneficial antibiotic-producing actinomycetes. This changes the soil to a disease-suppressive condition within a relatively short period. The probability that a particular beneficial microorganism will become predominant, even with organic farming or nature farming methods, will depend on the ecosystem and environmental conditions. It can take several hundred years for various species of higher and lower plants to interact and develop into a definable and stable ecosystem. Even if the population of a specific microorganism is increased through cultural and management practices, whether it will be beneficial to plants is another question. Thus, the likelihood of a beneficial, plant-associated microorganism becoming predominant under conservation-based farming systems is virtually impossible to predict. Moreover, it is very unlikely that the population of useful anaerobic microorganisms, which usually comprise only a small part of the soil microflora, would increase significantly even under natural farming conditions.

This information then emphasizes the need to develop methods for isolating and selecting different microor ganisms for their beneficial effects on soils and plants. The ultimate goal is to select microorganisms that are physiologically and ecologically compatible with each other and that can be introduced as mixed cultures into soil where their beneficial effects can be realised.

APPLICATION OF BENEFICIAL AND EFFECTIVE MICROORGANISMS: FUNDAMENTAL CONSIDERATIONS

Microorganisms are utilized in agriculture for various purposes; as important components of organic amen dments and composts, as legume inoculants for biological nitrogen fixation as a means of suppressing insects and plant diseases to Improve crop quality and yields, and for reduction of labour. All of these are closely related to each other. An important consideration in the application of beneficial microorganisms to soils is the enhancement of their synergistic effects. This is difficult to accomplish if these microorganisms are applied to achieve symptomatic therapy, as in the case of chemical fertilizers and pesticides.

If cultures of beneficial microorganisms are to be effective after inoculation into soil, it is important that their initial populations be at a certain critical threshold level. This helps to ensure that the amount of bioactive substances produced by them will be sufficient to achieve the desired positive effects on crop production and/or crop protection. If these conditions are not met, the introduced microorganisms, no matter how useful they are, will have little if any effect. At present, there are no chemical tests that can predict the probability of a particular soil-inoculated microorganism to achieve a desired result. The most reliable approach is to inoculate the beneficial microorganism into soil as part of a mixed culture, and at a sufficiently high inoculum density to maximize the probability of its adaptation to enviro nmental and ecological conditions.

The application of beneficial microorganisms to soil can help to define the structure and establishment of natural ecosystems. The greater the diversity of the cultivated plants that are grown and the more chemically complex the biomass, the greater the diversity of the soil microflora as to their types, numbers and activities.

The application of a wide range of different organic amen dments to soils can also help to ensure a greater microbial diversity. For example, combinations of various crop residues, animal manures, green manures, and municipal wastes applied periodically to soil will provide a higher level of microbial diversity than when only one of these materials is applied. The reason for this is that each of these organic materials has its own unique indigenous microflora which can greatly affect the resident soil microflora after they are applied, at least for a limited period.

SOIL CONTAMINATION ON SOIL ORGANISMS (DIAGNOSIS)

LABORATORY AND FIELD BIOASSAYS

The main characteristic of laboratory bioassays is that the potential toxicity of samples of field soil is studied in the laboratory using laboratory-bred test organisms. So, in such a test, well known organisms having similar characteristics are used, and the test can be performed under controlled conditions ruling out other possible disturbing influences. The advantage of bioassays is that they provide a direct indication of the toxicity of a specific soil, and integrate the effect of all substances present. The disadvantage is that the specific chemicals causing the observed effects cannot be identified. Bioassays can also be applied to determine the bioavailability of pollutants in soils as an indication of the potential risk for higher trophic levels.

Laboratory bioassays with earthworms *(Eisenia fetida)* and higher plants *(Cyperus esculentus)* have been described by Marquenie and Simmers and Van Gestel *et al.* the latter authors used these bioassays to study the influence of soil clean-up on the bioavailability of metals. In these methods, cylinders with a diameter of about 18 cm are filled with a 20 cm layer of soil. The cylinders have a perforated bottom and are placed in a dish filled with water. After one week of preincubation, test organisms are introduced. After 4 weeks, earthworms are sorted out of the soil; after incubation on wet filter paper for 24-48 hours to void the gut, they are analysed for metal content. Plants are harvested, and shoots are analysed. Van Gestel *et al.*described similar bioassays, using lettuce *(Lactuca sativa)* and radish *(Raphanus sativus)*, to determine metal bioavailability in soils. For these organisms, smaller amounts of soil are needed (about 400 mg), and as in the bioassays using *C. esculentus,* some nutrient solution was added to the soils to stimulate plant growth.

Bioassays using the earthworm species Lumbricus terrestris and E. fetida have been described by Menzie *et al.* They studied survival of the earthworms after exposure to contaminated soil for 28 and 14 days, respectively, and also determined the uptake of some selected chemicals in the earthworms.

Menzie *et al.*and Callahan *et al.* studied the potential risk of contaminated soils using *in situ* bioassays with earthworms. Contaminated soil was placed in plastic buckets placed in the ground from which the soil was taken. The buckets were constructed to allow for free exchange of air and water. Adult earthworms of the species *Lumbricus terrestris* were placed in the buckets, and observations were made after 1 and 7 days. Survival and morbidity (burrowing, coiling, shortening, swelling, lesions) were the test parameters, and earthworms (including gut content) were analysed to determine bioaccumulation of the main pollutants. This bioassay proved to give a good indication of the possible risk of polluted soil.

Two observations merit comment. First, the bioassays last only 7 days, which might be too short to allow for an equilibrium in the uptake of highly lipophilic chemicals. Thus, uptake of these chemicals by the earthworms may be underestimated. Second, uptake might be misjudged, because of the presence of contaminated soil in the earthworm gut. From the authors' experience, the gut content of an earthworm may account for about 50 per cent of its dry weight.

Kopezski used small enclosures (3 cm long; 4.8 cm in diameter) to study the impact of acidification on population growth and decomposition activity of the collembola species *Folsomia candida* and *Heteromurus nitidus* in a forest soil. The enclosures were filled with 1 g of hazel leaf litter and 1g of wafers, and buried into the soil. By using cheese cloth for the bottom and top ends of the enclosures, free contact with the surrounding soil was ensured. After 6 months, samples were analysed. A significant correlation appeared between animal numbers and soil pH, with *H. nitidus* being most sensitive. Decomposition showed a somewhat weaker correlation with soil pH.

To study the effects of N-deposition on the interactions between soil fauna and microflora and the effects on mineralization in coniferous forest soils, lysimeters have been used by Berg and Verhoef. Intact soil cores with total soil fauna or absence of mesofauna are treated with three $(NH_4)_2SO_4$ concentrations (10, 50, and 200 kg *N* per ha per g). The lysimeters are defaunated by means of microwave treatment. Before installation, they are incubated with a soil-spore suspension. Faunal groups extracted from the soil are added to the different treatments.

Migration of fauna between the lysimeter and the surrounding soil can occur through holes in the lysimeter. The lysimeters are covered by a gauze lid and covered just above the top by a plexiglass roof preventing rain input. The lysimeters are watered every two weeks by hand.

MUTAGENICITY TESTS

Besides the bioassays using invertebrates or higher plants, special methods may be necessary to assess the potential risk of contaminated soils. Some compounds present in soil as contaminants (*e.g.*, PAHs) are known mutagens, and their effects can be assessed by genotoxicity tests developed for drinking water, surface water, and sediments. These tests may be applicable for industrially contaminated sites, waste disposals, or sewage sludge-amended soils.

The usual approach to assess mutagenicity is to record the number of revertants in a Salmonella thyphimurium strain, plated with the test substance on a histidine deficient medium. To include those mutagens that require metabolic activation, a rat liver microsome suspension is added. In addition to this test, several other procedures have been proposed, which are often more sensitive than this test.` `

To apply mutagenicity tests to contaminated soils, an extract must be obtained. The extraction may be crucial to the validity of the results: extractions with solvents such as methanol or dimethylsulphoxide often induce a stronger mutagenic response in the Salmonella test, compared to water leachates; this difference in potency is especially true for superficial soil horizons.

The ecological relevance of mutagenicity test results for soils is difficult to evaluate, as this field of research is still underdeveloped. Mutagenicity is detected not only at contaminated sites but also in uncontaminated soil and often bears no clear relationship with the levels of known chemical mutagens in soil.

Field Studies (Biomonitoring)

Biomonitoring studies in soil often deal with the study of the effect of chemical stress on the structure and function of entire ecosystems. or at least at the community level. For that purpose. studies are performed to determine the impact on litter decomposition in forest ecosystems or on communities of soil mesofauna. Foodweb models for soil organisms have been constructed for agro-ecosystems. and are developed for a coniferous forest soil. based on stratified litterbag experiments. With these models. effects of excessive N-input on the soil ecosystem can be estimated. No standardized guidelines exist for such studies.

Also certain species of organisms can be selected. and followed in time to detect certain deviations that may be due to the impact of chemical contamination, a process called biomonitoring. Isopods are for instance. excellent bioindicators for metal contamination. because of their ability to concentrate metals in their body tissues.

Tolsma *et al*. recommended inclusion of primary producers (the plant species Urtica dioica and Holcus lanatus), detritivores (earthworms and isopods), and carnivores (mice or moles) in a biomonitoring system for the terrestrial environment. These organisms were selected because of their capability to accumulate metals. P AH. and organochlorine pesticides.

6

Basic Plant Nutrition

The living plant depends on a number of basic factors for normal growth:

- Light
- Air
- Water
- Nutrients
- Physical support

Soil plays an important role in all these factors except for light. If any of these basic factors are limiting, plant growth will be reduced or the life cycle may not be completed this is called the principle of limiting factors. In other words, plant growth potential is limited by the factor in shortest supply.

Yield may be reduced when one nutrient reaches excessive levels that cause toxicity, so the proper balance of nutrients is important. Also, other factors such as improper management or pests can lower yield. Therefore, a systems approach is necessary to integrate all the factors in the best combination to achieve the most economic yield.

ESSENTIAL PLANT NUTRIENTS

The plant mineral composition does not simply reflect the elements needed for growth. Plants can selectively absorb required elements for their growth. But they also can take up elements not needed for growth.

The terms essential plant nutrient or essential mineral element were formed to describe the minerals needed by plants to grow and complete life cycles.

Essential plant nutrients must be directly involved in some aspect of the plant metabolism such as structural material, enzymes or hormones, and they must not be totally replaceable by another mineral element. For higher plants such as canola, there are 14 essential nutrients (besides CO_2, oxygen and water).

Table. Essential Plant Nutrients

Macronutrients	N,	P,	K,	S,	Mg,	Ca	
Micronutrients	Fe,	Mn,		Zn,	Cu,	B,	Mo, Cl and Ni

Table also indicates that plant nutrients are often classified by the relative amounts needed. Macronutrients are needed in large amounts relative to micronutrients. Table shows the relative amounts of nutrients contained in a typical canola crop. Some nutrients can be accumulated in plants much higher than necessary for growth.

Table. Approximate Amounts of Nutrients in the Above- Ground Portion of a 1,960 kg/ha (35 bu/ac) Canola Crop

Element	kg/ha	lb/ac
Nitrogen (N)	112–134	100–120
Phosphorus (P)	1–28	15–25*
Potassium (K)	67–134	60–120*
Sulphur (S)	22–28	20–25
Calcium (Ca)	45–67	40–60
Magnesium (Mg)	13–20	12–18
Iron (Fe)	~1	~1
Chlorine (Cl)	~0.8	~0.7
Manganese (Mn)	~0.2	~0.2
Zinc (Zn)	~0.2	~0.2
Boron (B)	~0.2	~0.2
Copper (Cu)	~0.7	~0.06
Nickel (Ni)	~0.004	~0.004
Molybdenum (Mo)	~0.004	~0.004

Note:

* P X 2.3=P_2O_5; K X 1.2=K_2O

**Crop uptake of nutrients is greatly affected by conditions in the soil or weather (dry, wet, cold, compaction, nutrient imbalances, salinity, etc.).

GENERAL NUTRIENT UPTAKE

The following discussion outlines nutrient movement into and through the canola plant. The level of most nutrients in the plant sap is much higher than in the water surrounding the roots. For example, a typical N content in a canola plant at the rosette stage would be 5 to 6% N, whereas a fertile soil in the spring would contain about 0.0002% N in a plant available form on a dry weight basis. The N level in the soil solution would be in the range of 0.00002%. Therefore, plant nutrient uptake must be highly selective.

Nutrient uptake begins when plant available forms move from the soil water through pores in the root skin (exodermis) into the free space of the roots. This free space comprises about 5 to 10% of the root internal volume. This movement is a passive process (doesn't require energy from the plant) driven either by diffusion (movement due to differences in concentration) or mass flow (simply carried by water flowing into the roots). The movement is selective since pores into the free space act as a size filter. Many nutrient ion diameters are much smaller than the pores. For example, potassium and calcium are only 10 to 20% of the pore size, and have easy access to the free space. Large diameter substances such as metal chelates, viruses and fungi are restricted from entry by the small pore size.

As plant roots grow, the soil volume and surface area explored increases, which increases the capacity for nutrient absorption. In addition, roots possess a cation exchange capacity (CEC) due to negative charges in cell walls. This root CEC attracts positive ions (cations) like ammonium (NH_4+) but repels negative ions (anions) such as nitrate (NO_3-). (For details on CEC, see section âœSoil Properties that Affect Plant Nutrition.)

After entry into the free space, nutrients move into the cell interior by crossing a plasma membrane found on the inside of cell walls. Another similar membrane is found surrounding a large central storage compartment (vacuole) that usually fills more than 80% of the total cell volume. The plasma and vacuole membranes are effective barriers and are the main sites for nutrient uptake selectivity. These membranes contain carrier systems or ion pumps that transport certain nutrients. Such systems are called active since they require energy from the plant to work. This energy demand for ion uptake by roots is considerable, taking up to 1/3 of the energy during rapid growth. The energy for root activity arises from respiration, which requires carbohydrates and oxygen. This explains why nutrient uptake often stops in flooded soils there is a lack of oxygen.

Some active uptake systems are constant while others have a rate that can be regulated. As the plant level of nutrients and related compounds increases, the root uptake rate can decrease (negative feedback). In contrast, as plants build tissue, the level of nutrient âœ building blocks decreases, and the roots are signalled to increase the uptake rate (positive feedback) for nutrients. Passive ion channels through the membranes allow for selective nutrient movement.

The selectivity of the various transport systems across the membranes is not absolute. There is often competition between ions of similar size and charge. For example, chloride (Cl-) competes with nitrate (NO_3-). This competition between Cl– and NO_3 - is important in certain saline soils with Cl- as a major component of the salt. Most prairie soils contain salt with sulphate as the main anion.

Since cation and anion uptake are regulated differently, plants must be able to compensate for differences in electrical charges that arise from disproportionate uptake of cations and anions. Plant cells maintain a pH in the range 7.3 to 7.6 by either releasing or consuming hydrogen cations (H+), which is achieved by formation or removal of organic acids. The nutrient journey continues in a path from cell to cell through tiny connecting tubes (plasmodesmata), although some nutrients can continue to move between cells through the free space. The next barrier occurs at the waxy layer (Casparian band) that surrounds the central vascular tissue (phloem and xylem). The phloem and xylem are special tissues that act like highways for nutrient transport from roots to leaves (xylem) and from leaves to growing points and roots (phloem). In young root tips, the Casparian strip is not well formed and thus is an incomplete barrier.

The mechanism how ions pass through the Casparian strip and into the xylem (âœxylem loading) is not well understood. There is probably a combination of active (ion pumps) and passive channels for ion movement into the xylem. Xylem loading is regulated separately from root uptake, thus creating a control system for nutrient movement. Nutrients are carried by water up through the xylem. Water flows up through xylem tissues due to a suction-like force created when water evaporates from the leaves, and from slight pressure produced by roots. Once inside the xylem sap, nutrients can be unloaded and reloaded before reaching the end growing points.

Once nutrients reach their targets, there often is considerable recycling, especially for the mobile nutrients such as N. For example, a normal feature of plants appears to be simultaneous import and export of nutrients from leaves. This dynamic nutrient cycling is termed remobilization or retranslocation. In young vegetative plants, nutrient recycling occurs from the mature leaves to roots and young leaves through the phloem. The remobilization ability of different nutrients affects where deficiency symptoms occur. Deficiency symptoms of mobile nutrients such as N will first appear in old tissues. In contrast, deficiency symptoms of nutrients with limited mobility such as sulphur and copper will occur in young tissue, and can hinder flower/ seed development.

Nutrient remobilization is particularly important when seeds are forming. At this stage, mobile nutrients are being exported from ageing leaves while nutrient imports are decreasing. Also, root activity and nutrient uptake generally decrease by this stage due to drying soils, nutrient depletion in the soil and a relative shift in the energy supply from roots to developing pods and seeds. Plant parts with a strong energy or nutrient demand are called sinks. As a result, old leaves are sacrificed to supply pod and seed growth. Mobile nutrients in the seed have mostly been transferred from other plant tissue"in canola the sources are pods, stems and leaves.

Roots are not the only sites where nutrient uptake can occur. Some nutrients can be absorbed by leaves and other above ground plant parts. Nutrients in the gas form NH_3 (ammonia), NO_2 (nitrogen dioxide) and SO_2 (sulphur dioxide) can enter leaves through leaf pores (stomata) and then be changed into organic forms. These gases are major air pollution components and in some areas contribute considerably to plant nutrition. In areas with intensive livestock operations, NH_3 uptake can contribute 10 to 20% of the nitrogen for adjacent crops. SO_2 is readily absorbed by leaves. In a European field experiment, almost half of the total sulphur (S) taken up by vegetative rapeseed came from atmospheric S compounds, probably SO_2. This may partly explain why S deficiencies have increased in western Canada after environmental regulations enforced cleanup of S emissions from gas plants.

SOIL PROPERTIES THAT AFFECT PLANT NUTRITION

Soil is a complex mixture of non-living substances (minerals, organic matter, gases and liquids) and living organisms (bacteria, fungi, insects, worms, etc.). These factors influence soil fertility either directly or indirectly.

Soil solids consist of mineral particles, organic matter in varying stages of decomposition and living organisms. Solids make up about half the soil volume, while water and gases make up the other half in the pore space.

Soil mineral particles vary widely in size and are classified by size:

- Rocks are larger than 2 mm in diameter (0.08")
- Sand particles range from 0.05 to 2 mm (0.002 to 0.08") in diameter
- Silt particles range from 0.002 to 0.05 mm (0.00008 to 0.002") in diameter
- Clay particles are smaller than 0.002 mm in diameter"(0.00008") in diameter

These particles are made from various mineral types with different elemental composition, which affects weathering processes and thus the release of certain nutrients. Two soils with identical texture could be drastically different in fertility due to differences in mineral composition. Potassium is an example of a plant nutrient whose supply arises from mineral weathering in soil.

The soil colloidal fraction refers to microscopic particles of clay and organic matter. The surface of the colloidal fraction is where most soil chemical reactions occur and it is very important in nutrient supply.

The proportion of sand, silt and clay determines the texture of a soil. Soil texture is grouped into five or more classes. The texture influences fertility by affecting moisture holding capacity, air exchange and the CEC. Adequate moisture is key to fertilizer response and potential yield for canola in western Canada. The CEC is an important property that influences the soil storage of many plant nutrients. Most nutrients are present in the soil water as positively charged cations. A few are negatively charged anions. The CEC indicates a soils ability to hold or store cations. Prairie soil particles typically have a negative charge. The process of electrical attraction that holds cations to negative surfaces of soil colloids is called adsorption (not absorption). The cations are not permanently stuck to the colloidal surface and can be exchanged with other cations. With time certain cations may become âœ fixed into forms that are not easily removed from the exchange complexes. Adsorbed cations are not removed by water moving through the soil and can be accessed by plant roots. Cations with a higher positive charge (for example Ca^{+2}) are held more tightly than those with a lower charge (for example K+). Soil negative charges arise due to substitutions in the mineral crystals by elements with smaller positive charge, and due to reactions at the edges. Organic particles also contain a significant number of negative charges. The total particle surface area in a soil increases as particle sizes get smaller. Therefore, a soil high in clay has a much greater surface area than a sandy soil.

A high clay soil also has a bearing on the surface area and negative charge. Clay minerals are microscopic layers of aluminium and silicon crystals formed by weathering of other minerals. Thus clays are called secondary minerals. The type of clay depends on the original minerals and the weathering extent. Fairly âœ young clays common in western Canada (such as montmorillonite) have a 2:1 arrangement of silica:alumina crystal sheets, while older clays have a 1:1 arrangement. Generally, 2:1 clays have 10 to 100 times more surface area, negative charges and, consequently, a higher CEC than 1:1 clays. Organic colloids have 10 to 100 times more negative charges and higher CECs than the 1:1 clays. Therefore, soil organic matter levels greatly influence the CEC.

The CEC strongly influences soil fertility. A higher CEC means that more cations, including plant nutrients, can be loosely stored in a plant available form, giving the plant a greater pool of nutrients to draw from. Since most cations are not highly soluble, only small quantities can be dissolved in the soil solution at one time.

The CEC soil property allows a reservoir of nutrients to be stored then released to plant roots. This continuous replenish ment of nutrients in soil water is very important for several nutrients, including potassium. A high CEC also means that fewer cations will be lost through leaching out of the root zone. Since soils are predominantly negatively charged, anions [such as nitrate (NO_3^-) and sulphate (SO_4^{-2})] are repelled by soil colloids and tend to stay in the soil water. They will flow with water and are potentially subject to leaching loss. Soil organic matter (OM) plays an important role in soil fertility as a plant nutrient storehouse. Not only does OM adsorb many cations due to a high CEC, it also stores nutrients as part of its structure. As the OM is decomposed by soil microbes, nutrients are released from the organic structure into plant available forms"this process is called mineralization.

Mineralization from OM is the primary natural source of plant available N and S in prairie soils, and also influences P availability. Minera lization of individual nutrients will be described in later sections. Soil OM also plays a secondary role in soil fertility by improving physical properties such as water holding capacity, infiltration, aggregation (tilth) and buffering pH.

SOIL MICROORGANISMS ON PLANT HEALTH AND NUTRITION

Soil microorganisms, sometimes spelled as soil micro-organisms, are a very important element of healthy soil. Knowing what microbes in soil eat, the conditions they thrive in and the temperatures that they are most active in is important in organic gardening and organic lawn care. From a practical standpoint, it boils down to organic matter, but not just any organic matter. These facts below will help you plan your activities around the time they are most beneficial. Below is a partial list of important functions they perform.

Soil microorganisms are responsible for:

- Ansforming raw elements from one chemical form to another. Important nutrients in the soil are released by microbial activity are Nitrogen, Phosphorus, Sulfur, Iron and others.
- Breaking down soil organic matter into a form useful to plants. This increases soil fertility by making nutrients available and raising CEC levels.
- Degradation of pesticides and other chemicals found in the soil.
- Suppression of pathogenic microorganisms that cause diseases. The pathogens themselves are part of this group, but are highly outnumbered by beneficial microbes.

TYPES OF MICROORGANISMS IN SOIL

There are several types of microorganisms in soil that benefit plants. Together they make up an immense population of living organisms. One teaspoon of soil may contain millions of various types. Below is a list of common soil microor-ganisms found throughout the world.

- *Bacteria*: Small, single cell organisms that make up the single most abundant type of microbe. They have a very wide range of conditions that they live in from the artic wastelands to the steaming waters of volcanic hot springs. In soils, they multiply rapidly under the proper conditions. When conditions are wrong for one species, it is right for another. This is not always a good thing since a balance is what is required.
- *Fungi:* The largest microbe group in terms of mass. Some fungi are beneficial, called mycorrhiza, that form a symbiotic relationship with plant roots, either externally or internally. Within the fungi group are pathogen fungi. These are disease causing fungi, some of which can be quite devastating to plants.
- *Protozoas*: Small single cell microbes that feed on bacteria.
- *Actinomycetes:* Necessary for the breakdown of certain components in organic matter.
- *Algae:* Beneficial groups such as blue-green algae, yellow-green algae and diatoms. Some of these can produce their own energy through photosynthesis.

Soil microorganisms are living, breathing organisms and, therefore, need to eat. They compete with plants for nutrients including Nitrogen, Phosphorus,

Potassium and micronutrients as well. They also consume amino acids, vitamins, and other soil compounds. Their nutrients are primarily derived from the organic matter they feed upon. The benefit is that they also give back or perform other functions that benefit higher plant life.

ORGANIC MATTER

It is a variety of natural substances including decomposed leaves, grass clipping, shed roots, wood chips, etc. Humus (well decomposed organic matter) is the richest source for plant growth. Organic matter comes in many different nutrient levels, especially Nitrogen. While soil microbes need carbon (C) to live, they also need the nitrogen contained in organic matter. Therefore, the Carbon to Nitrogen ratio (C:N) is very important.

Problems with Low Nitrogen Organic Matter

Organic matter low in Nitrogen will also have a slower breakdown rate than organic matter with higher nitrogen levels. The microbes will consume the Nitrogen element first and the grass will get what is left over. So organic matter high in carbon and low in Nitrogen will provide little nitrogen to the grass. However, if another N source is applied over the organic matter, it will speed up the decomposition.

Therefore, the rule of thumb is to choose an organic matter with higher levels of nitrogen. Anything lower than four per cent (4%) Nitrogen with high Carbon content should be considered a soil amendment and not a fertilizer.

How Temperature Affects Soil Microbe Activity

Soil Microbe activity is dependent on soil tempe ratures. For simplicity, all essential soil microbes are classified into the three different temperature ranges they are most active in.

1. *Psychrophiles*: Active in temperatures less than 68 degrees.
2. *Mesophiles*: Active in temperatures between 77 degrees and 95 degrees. This makes up the largest group of soil microbes and the range most activity charts are based on.
3. *Thermopholes*: Active in temperatures from 115 degrees to 150 degrees. From a plant and landscape view, this group will rarely apply.

Since the primary group contains Mesophiles, this has a great influence on the degree of soil microbe activity. Areas where the temperatures are warm most of the year, organic matter can be consumed very quickly. Tropical rainforests are so lush in part because of consistently warm temperatures, which promote fast breakdown of organic matter and the release of nutrients into the soil.

Cooler areas, such as Canada and parts of Europe, that have extended periods of cool weather below 77 degrees will benefit far less from the additions of organic matter. This is because far fewer microbes are active in that temperature range. It is possible to build unhealthy levels of organic matter if you follow the example of those in warmer climates.

The scientific rule is this: With every 18 degree rise in temperature, from 32 degrees to 95 degrees, there is a 1.5 to 3.0 % increase in microbial activity. (Carrol/Waddington/Rieke) Remember, the food availabity to microbes, the quality of organic matter, soil types, pH level, per cent of Nitrogen, etc. will also have an effect on microbial activity level.

SOIL PH FACTOR

Most soil microorganisms can tolerate a wide range of soil levels. However, bacteria favours a neutral to slightly alkaline soil up to 8.0. When pH drops below 6, fungi begin to dominate as bacteria finds it less favourable.

SOIL MOISTURE

Just as temperature levels stimulate different soil microbes, so does soil moisture. Some are obvious. Persistent, damp conditions with heavy shade will promote the growth of algae while hindering microbes that thrive in sunny locations. Proper lawn watering requires deep watering so that the soil is wet 4 inches deep. Shallow watering means only the surface is wet. It drys out quickly and can greatly hinder soil microbes.

OXYGEN LEVELS NECESSARY FOR HEALTHY MICROBES

There is a balance to everything, including oxygen in soil. Compacted soil will have less oxygen and less water holding capacity. Clay soil consists of extremely tiny particles, even smaller than silt. Clay with proper structure can have sufficient oxygen, but it can also compact easily. Since soil microorganisms consume oxygen, but low oxygen soils will quickly deplete what oxygen it has and lower the soil microorganism levels. In lawns, deeply water to a level of 4 inches deep and allow it to dry before watering again. The deeper soil will remain moist at sufficient levels. Shallow watering means when the surface moisture is gone there is no moisture deeper to support a healthy microbe population.

DECOMPOSITION

Well decomposed organic matter is the oldest form of biostimulant and very effective. Modern advancements allow us to apply specific bacteria and other ingredients necessary for healthy plants. Other than organic matter, humic acid is a well performing biostimulant for lawns and gardens that has shown to stimulate soil micro organisms.

Humic acid is known to:

- Enhance soil nutrient content.
- Increase CEC.

- Stimulate and increase microbial numbers.
- Increases soil moisture holding capacity.
- Improves soil structure.

Soil microorganisms are one of the most important elements of a healthy soil. A good lawn care programme will take advantage of the many benefits of soil microbes and the biostimulants that encourage them.

TRANSFORMATIONS THAT SUPPLY PLANT AVAILABLE NITROGEN

Soil organic matter is a major N reservoir containing several thousand pounds of organic N per acre. This large organic N storehouse needs to be decomposed by soil microbes before becoming available for root uptake. The decomposition process is called mineralization.

The decomposition rate is fairly slow and variable, ranging from about 0.4 to 5% per year. As shown in Figure, organic N is first slowly changed to ammonium by bacteria. Then different bacteria rapidly change the ammonium to nitrate, in a two-step process called nitrification.

The ammonium is first changed to nitrite (NO_2 -), then to nitrate. Under normal soil conditions, the bacterial oxidation of ammonium to nitrite is much slower than nitrite to nitrate. Therefore, very little nitrite is normally found in soil. This is fortunate since nitrite is toxic to plants. The result of these rate differences is that most of the plant available N in the soil tends to be nitrate. This is also why most soil testing labs analyse soil for nitrate and dont include ammonium.

Since N transformations result from soil microbial activity, soil conditions such as temperature, moisture and acidity will strongly affect the rates at which these processes occur. If soil is cold, saturated or very acidic, then mineralization and other microbial activities will proceed slowly. Cultivation stimulates organic matter decomposition and mineralization of N by improving aeration and physical mixing that gives soil microbes access to new organic matter supplies.

There are several sources of plant available N to the soil system, in addition to soil organic matter decomposition and fertilizer additions (Figure). The atmosphere contains 78% N_2 and thus is a huge source of N. But it first must be changed to plant available forms through biological or industrial processes called N_2 fixation. Biological fixation of atmospheric N_2 is performed by certain bacteria or blue-green algae species.

Biological N fixation in soil falls into three general types:

- Symbiosis with legumes
- Associative
- Free-living N fixing bacteria

Symbiotic N fixation is much larger relative to the other types. Canola is not a legume and cannot form the symbiosis with rhizobia to fix atmospheric

N. However, canola can benefit from residual N fixed by previous legume crops. Associative N fixation occurs when bacteria just inside the root or on the root surface use root exudates for energy to fix atmospheric N. The plant benefits indirectly when the bacteria dies and its N is released through mineralization. The results of many experiments in Russia and throughout the world on cereal inoculation with associative N fixing bacteria have shown varying and unpredictable responses. This suggests that the interactions between plants and the bacteria are complex, unstable and can vary greatly depending on genotypes.

Much less research has been done on canola inoculation with associative N fixing bacteria. The limited research does show that some strains will successfully colonize the roots of Brassica species but often do not significantly influence harvest dry weight or N accumulation.

Transformations that Reduce Plant Available Nitrogen

Several different mechanisms contribute to N loss from soil, and, therefore, lost opportunity for increasing yield. Understanding the conditions that promote such losses can be valuable in avoiding such conditions in the field and improving the N fertilizer efficiency.

Denitrification

One major N loss from soil occurs through a microbial process called denitrification. As shown in Figure, nitrate can be changed by certain bacteria to gases such as N_2O and N_2, which escape back to the atmosphere. These soil bacteria have the ability to switch their respiration from using oxygen to nitrate. Since respiration is more efficient using oxygen, these denitrifying bacteria will only switch to nitrate if oxygen is absent, such as in waterlogged soil. Therefore, denitrification becomes significant when soils become saturated. Other secondary factors that encourage denitrification include carbon availability (crop residues), warm soil, and neutral to alkaline pH.

Denitrification accounts for 10 to 50% of the available N losses in prairie soil. Research on the prairies has shown that considerable N denitrification losses occur during spring thaw. For example, research near Edmonton, AB found that 16 to 60% of annual denitrification loss occurred immediately following snowmelt. During spring thaw on the prairies, frozen subsoil is often a barrier for water drainage and the overlying thawed soil becomes saturated.

The second major period of denitrification loss occurs in late spring and early summer during rainfall events that cause soil saturation. Remember that denitrification occurs regardless of the nitrate source from fertilizer, manure or from decomposition. Effective N fertilizer management strives to avoid having large amounts of nitrate present during spring melts. Summerfallow is especially prone to large denitrification losses since large amounts of nitrate and moisture are stored during the fallow year, which increases the denitrification potential in the next spring thaw. Summerfallow also is a major contributor to leaching losses of nitrate.

Immobilization

Immobilization is the second major N transformation that reduces the plant available N supply. Figure indicates that soil bacteria may use either nitrate or ammonium for their own growth, temporarily tying up the N in the soil organic N storehouse.

Immobilization essentially is the reverse of mineralization, and occurs when residues with low N content (like cereal straw) are being decomposed. Since these residues don't contain enough N for the microbes to make their own protein, they need to use the nitrate and ammonium. Soil microbes thus compete with plant roots for the available N, and plant growth suffers when N supplies are inadequate for both microbial and plant growth needs. The poor crop growth in heavy chaff rows is due in part to immobilization of N and other nutrients by the decomposing microbes. One effective fertilization strategy is to place the fertilizer away from residues and thus avoid immobilization losses. The remaining transformations that reduce plant available N are usually relatively minor.

Table. Seedbed Utilization (SBU) of Various Openers*

	Spread Width of Fertilizer in Seed Row											
	2.5 cm (1") Disc or Knife			5 cm (2") Spoon or Hoe			7.6 cm (3") Sweep			10 cm (4") Sweep		
Row spacing cm	15	23	30	15	23	30	15	23	30	15	23	30
Row spacing "	6	9	12	6	9	12	6	9	12	6	9	12
SBU %	17	11	8	33	22	17	50	33	25	67	44	33

Note: *Although some openers also vertically spread seed and fertilizer, this is not considered in the table since seed should be placed at a consistent depth for uniform germination and emergence. The actual spread width varies with air flow, soil type, moisture, residue and speed, and should be checked under prevailing field conditions.

Leaching

Leaching of nitrate can occur since this form is not adsorbed to the soil and moves readily with soil water. Leaching losses can be significant in sandy soils in high rainfall areas or under summerfallow, but overall leaching probably contributes to less than 10% of the available N losses on the prairies.

To reduce leaching losses time fertil izer applications to avoid prolonged exposure to wet conditions, and consider band placement to delay the conversion to the vulnerable nitrate form.

Volatilization

Volatilization occurs when ammonia escapes from the soil to the atmosphere. Such losses happen in a variety of ways. One obvious loss occurs when anhydrous ammonia fertilizer is improperly applied (too shallow or into a too dry or wet soil). Broadcasting urea fertilizer on the surface without incorporation can also lead to significant volat ilization losses if significant rainfall (more than 6 mm (1/4") does not occur soon after application. All ammonium based fertilizers are subject to volatilization if broadcast on the surface of soils with high pH, surface lime salts, low soil organic matter, warm temperatures and dry conditions. To reduce volatilization loss, ensure proper fertilizer placement into the soil.

Weeds

Weeds can contribute to poor N fertilizer efficiency by competing with crops for uptake. The competitive ability of the crop for fertilizer uptake can be improved by placing the fertilizer near crop roots rather than broadcasting or random banding.

Erosion

Erosion of topsoil carries significant N and other nutrients away from the field. Use soil conservation techniques to minimize such losses.

A final minor loss mechanism occurs when ammonium is fixed into the crystal structure of certain clays. Some soils contain expanding type clays that allow ammonium to enter within the plates of the crystal structure and become fixed. Such ammonium trapped within the crystal lattice is held tightly and unavailable for root uptake.

ELEMENTS OF COMPLETE PLANT NUTRITION

The following is a brief guideline of the role of essential and beneficial mineral nutrients that are crucial for growth. Eliminate any one of these elements, and plants will display abnormalities of growth, deficiency symptoms, or may not reproduce normally.

MACRONUTRIENTS

Nitrogen: Is a major component of proteins, hormones, chlorophyll, vitamins and enzymes essential for plant life. Nitrogen metabolism is a major factor in stem and leaf growth (vegetative growth). Too much can delay flowering and fruiting. Deficiencies can reduce yields, cause yellowing of the leaves and stunt growth.

Phosphorus: Is necessary for seed germination, photosynthesis, protein formation and almost all aspects of growth and metabolism in plants. It is essential for flower and fruit formation. Low pH (<4) results in phosphate being chemically locked up in organic soils. Deficiency symptoms are purple stems and leaves; maturity and growth are retarded. Yields of fruit and flowers are poor. Premature drop of fruits and flowers may often occur. Phosphorus must be applied close to the plant's roots in order for the plant to utilize it. Large applications of phosphorus without adequate levels of zinc can cause a zinc deficiency.

Potassium: Is necessary for formation of sugars, starches, carbohydrates, protein synthesis and cell division in roots and other parts of the plant. It helps to adjust water balance, improves stem rigidity and cold hardiness, enhances flavour and colour on fruit and vegetable crops, increases the oil content of fruits and is important for leafy crops. Deficiencies result in low yields, mottled, spotted or curled leaves, scorched or burned look to leaves..

Sulfur: Is a structural component of amino acids, proteins, vitamins and enzymes and is essential to produce chlorophyll. It imparts flavour to many vegetables. Deficiencies show as light green leaves. Sulfur is readily lost by leaching from soils and should be applied with a nutrient formula. Some water supplies may contain Sulfur.

Magnesium: Is a critical structural component of the chlorophyll molecule and is necessary for functioning of plant enzymes to produce carbohydrates, sugars and fats. It is used for fruit and nut formation and essential for germination of seeds. Deficient plants appear chlorotic, show yellowing between veins of older leaves; leaves may droop. Magnesium is leached by watering and must be supplied when feeding. It can be applied as a foliar spray to correct deficiencies.

Calcium: Activates enzymes, is a structural component of cell walls, influences water movement in cells and is necessary for cell growth and division. Some plants must have calcium to take up nitrogen and other minerals. Calcium is easily leached. Calcium, once deposited in plant tissue, is immobile (non-translocatable) so there must be a constant supply for growth. Deficiency causes stunting of new growth in stems, flowers and roots. Symptoms range from distorted new growth to black spots on leaves and fruit. Yellow leaf margins may also appear.

MICRONUTRIENTS

Iron: Is necessary for many enzyme functions and as a catalyst for the synthesis of chlorophyll. It is essential for the young growing parts of plants. Deficiencies are pale leaf colour of young leaves followed by yellowing of leaves and large veins. Iron is lost by leaching and is held in the lower portions of the soil structure. Under conditions of high pH (alkaline) iron is rendered unavailable to plants. When soils are alkaline, iron may be abundant but unavailable. Applications of an acid nutrient formula containing iron chelates, held in soluble form, should correct the problem.

Manganese: Is involved in enzyme activity for photosynthesis, respiration, and nitrogen metabolism. Deficiency in young leaves may show a network of green veins on a light green background similar to an iron deficiency. In the advanced stages the light green parts become white, and leaves are shed. Brownish, black, or grayish spots may appear next to the veins. In neutral or alkaline soils plants often show deficiency symptoms. In highly acid soils, manganese may be available to the extent that it results in toxicity.

Boron: Is necessary for cell wall formation, membrane integrity, calcium uptake and may aid in the translocation of sugars. Boron affects at least 16 functions in plants. These functions include flowering, pollen germination, fruiting, cell division, water relationships and the movement of hormones. Boron must be available throughout the life of the plant. It is not translocated and is easily leached from soils. Deficiencies kill terminal buds leaving a rosette effect on the plant. Leaves are thick, curled and brittle. Fruits, tubers and roots are discolored, cracked and flecked with brown spots.

Zinc: Is a component of enzymes or a functional cofactor of a large number of enzymes including auxins (plant growth hormones). It is essential to carbohydrate metabolism, protein synthesis and internodal elongation (stem growth). Deficient plants have mottled leaves with irregular chlorotic areas. Zinc deficiency leads to iron deficiency causing similar symptoms. Deficiency occurs on eroded soils and is least available at a pH range of 5.5 - 7.0. Lowering the pH can render zinc more available to the point of toxicity.

Copper: Is concentrated in roots of plants and plays a part in nitrogen metabolism. It is a component of several enzymes and may be part of the enzyme systems that use carbohydrates and proteins. Deficiencies cause die back of the shoot tips, and terminal leaves develop brown spots. Copper is bound tightly in organic matter and may be deficient in highly organic soils. It is not readily lost from soil but may often be unavailable. Too much copper can cause toxicity.

Molybdenum: Is a structural component of the enzyme that reduces nitrates to ammonia. Without it, the synthesis of proteins is blocked and plant growth ceases. Root nodule (nitrogen fixing) bacteria also require it. Seeds may not form completely, and nitrogen deficiency may occur if plants are lacking molybdenum. Deficiency signs are pale green leaves with rolled or cupped margins.

Chlorine: Is involved in osmosis (movement of water or solutes in cells), the ionic balance necessary for plants to take up mineral elements and in photosynthesis. Deficiency symptoms include wilting, stubby roots, chlorosis (yellowing) and bronzing. Odors in some plants may be decreased. Chloride, the ionic form of chlorine used by plants, is usually found in soluble forms and is lost by leaching. Some plants may show signs of toxicity if levels are too high.

Nickel: Has just recently won the status as an essential trace element for plants according to the Agricultural Research Service Plant, Soil and Nutrition Laboratory in Ithaca, NY. It is required for the enzyme urease to break down urea to liberate the nitrogen into a usable form for plants. Nickel is required for iron absorption. Seeds need nickel in order to germinate. Plants grown without additional nickel will gradually reach a deficient level at about the time they mature and begin reproductive growth. If nickel is deficient plants may fail to produce viable seeds.

Sodium: Is involved in osmotic (water movement) and ionic balance in plants.

Cobalt: Is required for nitrogen fixation in legumes and in root nodules of nonlegumes. The demand for cobalt is much higher for nitrogen fixation than for ammonium nutrition. Deficient levels could result in nitrogen deficiency symptoms.

Silicon: Is found as a component of cell walls. Plants with supplies of soluble silicon produce stronger, tougher cell walls making them a mechanical barrier to piercing and sucking insects. This significantly enhances plant heat and drought tolerance. Foliar sprays of silicon have also shown benefits reducing populations of aphids on field crops. Tests have also found that silicon can be deposited by the plants at the site of infection by fungus to combat the penetration of the cell walls by the attacking fungus. Improved leaf erectness, stem strength and prevention or depression of iron and manganese toxicity have all been noted as effects from silicon. Silicon has not been determined essential for all plants but may be beneficial for many. Unlike animals (which obrtain their food from what they eat) plants obtain their nutrition from the soil and atmosphere. Using sunlight as an energy source, plants are capable of making all the organic macromolecules they need by modifications of the sugars they form by photosynthesis. However, plants must take up various minerals through their root systems for use.

A (PLANT) BALANCED DIET

Carbon, Hydrogen, and Oxygen are considered the essential elements. Nitrogen, Potassium, and Phosphorous are obtained from the soil and are the primary macronutrients. Calcium, Magnesium, and Sulfur are the secondary macronutrients needed in lesser quantity. The micronutrients, needed in very small quantities and toxic in large quantities, include Iron, Manganese, Copper, Zinc, Boron, and Chlorine. A complete fertilizer provides all three primary macronutrients and some of the secondary and micro-nutrients. The label of the fertilizer will list numbers, for example 5–10–5, which refer to the per cent by weight of the primary macronutrients.

SOILS PLAY A ROLE

Soil is weathered, decomposed rock and mineral (geological) fragments mixed with air and water. Fertile soil contains the nutrients in a readily available form that plants require for growth. The roots of the plant act as miners moving through the soil and bringing needed minerals into the plant roots.

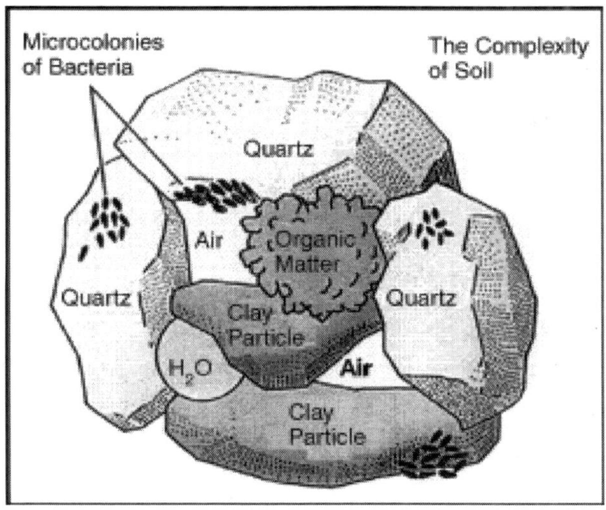

Plants use these minerals in:

- Structural components in carbohydrates and proteins
- Organic molecules used in metabolism, such as the Magnesium in chlorophyll and the Phosphorous found in ATP
- Enzyme activators like potassium, which activates possibly fifty enzymes
- Maintaining osmotic balance

Mycorrhizae, Bacteria, and Minerals

Plants need nitrogen for many important biological molecules including nucleotides and proteins. However, the nitrogen in the atmosphere is not in a form that plants can utilize.

Many plants have a symbiotic relationship with bacteria growing in their roots: organic nitrogen as rent for space to live.

These plants tend to have root nodules in which the nitrogen-fixing bacteria live.

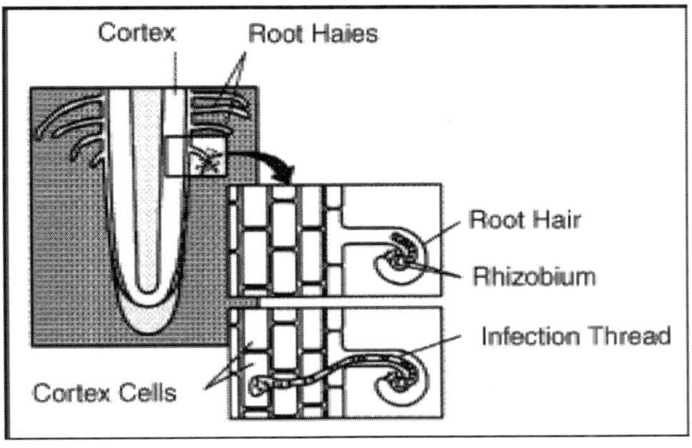

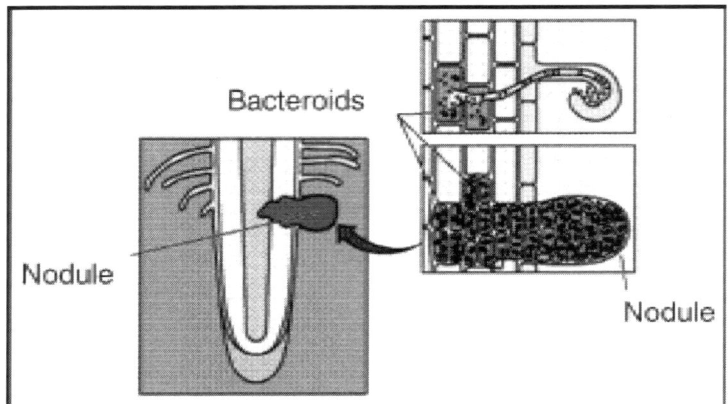

All the nitrogen in living systems was at one time processed by these bacteria, who took atmospheric nitrogen (N_2) and modified it to a form that living things could utilize (such as NO_3 or NO_4; or even as ammonia, NH_3 in the example shown below).

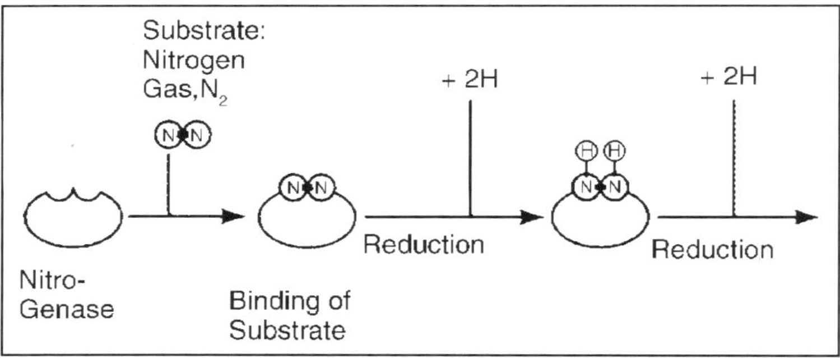

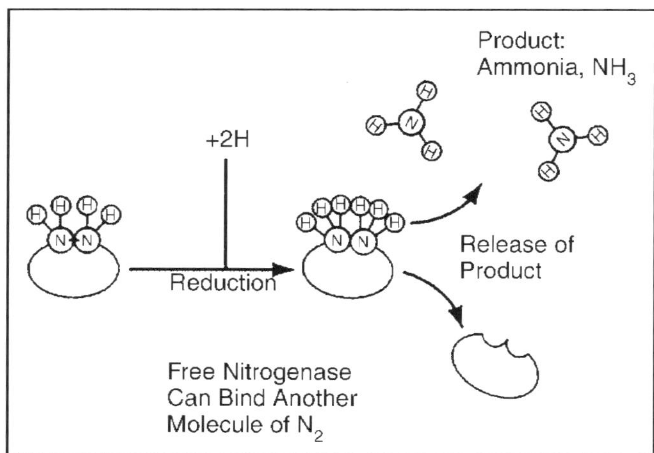

Not all bacteria utilize the above route of nitrogen fixation. Many that live free in the soil, utilize other chemical pathways.

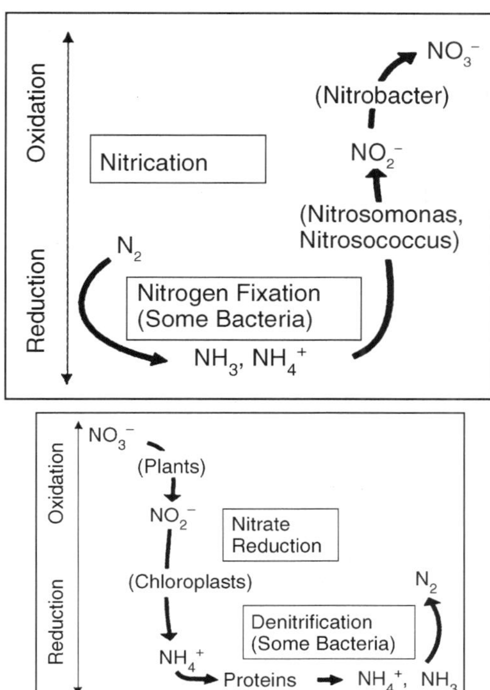

Roots have extensions of the root epidemal cells known as root hairs. While root hairs greatly enhance the surface area (hence absorbtion surface), the addition of symbiotic mycorrhizae fungi vastly increases the area of the root for absorbing water and minerals from the soil.

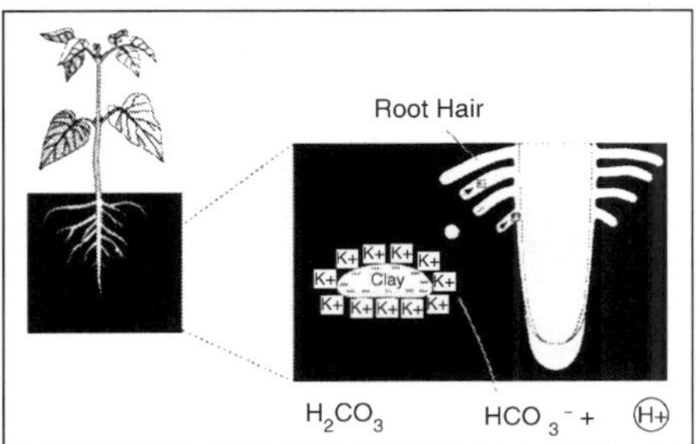

WATER AND MINERAL UPTAKE

Animals have a circulatory system that transports fluids, chemicals, and nutrients around within the animal body. Some plants have an analogous system: the vascular system in vascular plants; trumpet hyphae in bryophytes.

Root hairs are thin-walled extensions of the epidermal cells in roots. They provide increased surface area and thus more efficient absorption of water and minerals. Water and dissolved mineral nutrients enter the plant via two routes.

Water and selected solutes pass through only the cell membrane of the epidermis of the root hair and then through plasmodesmata on every cell until they reach the xylem: intracellular route (apoplastic). Water and solutes enter the cell wall of the root hair and pass between the wall and plasma membrane until the encounter the endodermis, a layer of cells that they must pass through to enter the xylem: extracellular route (symplastic).

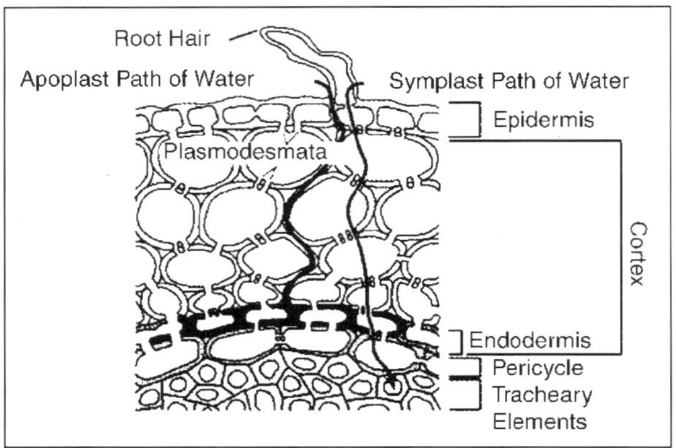

The endodermis has a strip of water-proof material (containing suberin) known as the Casparian strip that forces water through the endodermal cell and in such a way regulates the amount of water getting to the xylem. Only when water concentrations inside the endodermal cell fall below that of the cortex parenchyma cells does water flow into the endodermis and on into the xylem.

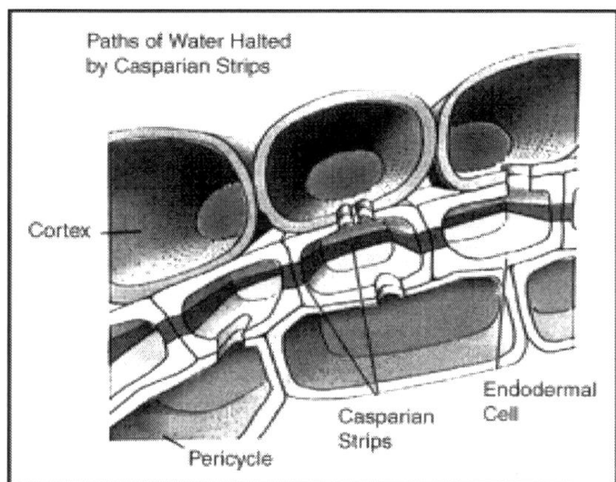

XYLEM AND TRANSPORT

Xylem is the water transporting tissue in plants that is dead when it reaches functional maturity. Tracheids are long, tapered cells of xylem that have end plates on the cells that contain a great many crossbars. Tracheid walls are festooned with pits. Vessels, an improved form of tracheid, have no (or very few) obstructions (crossbars) on the top or bottom of the cell. The functional diameter of vessels is greater than that of tracheids. Water is pulled up the xylem by the force of transpiration, water loss from leaves. Mature corn plants can each transpire four gallons of water per week. Transpiration rates in arid-region plants can be even higher. Water molecules are hydrogen bonded to each other. Water lost from the leaves causes diffusion of additional water molecules out of the leaf vein xylem, creating a tug on water molecules along the water columns within the xylem. This "tug" causes water molecules to rise up from the roots to eventually the leaves. The loss of water from the root xylem allows additional water to pass throught the endodermis into the root xylem.

Cohesion is the ability of molecules of the same kind to stick together. Water molecules are polar, having slight positive and negative sides, which causes their cohesion. Inside the xylem, water molecules are in a long chain extending from the roots to the leaves. Adhesion is the tendency of molecules of different kinds to stick together. Water sticks to the cellulose molecules in the walls of the xylem, counteracting the force of gravity and aiding the rise of water within the xylem.

COHESION-ADHESION THEORY

Transpiration exerts a pull on the water column within the xylem. The lost water molecules are replaced by water from the xylem of the leaf veins, causing a tug on water in the xylem. Adhesion of water to the cell walls of the xylem facilitates movement of water upward within the xylem. This combination of cohesive and adhesive forces is referred to as the Cohesion-Adhesion Theory.

GUARD CELLS REGULATE TRANSPIRATION

In most environments, the water concentration outside the leaf is less than that inside the leaf, causing a loss of water through openings in the leaf known as stomata (singular = stoma). Guard cells are crescent-shaped cells of the epidermis that flank the stoma and regulate the size of the opening.

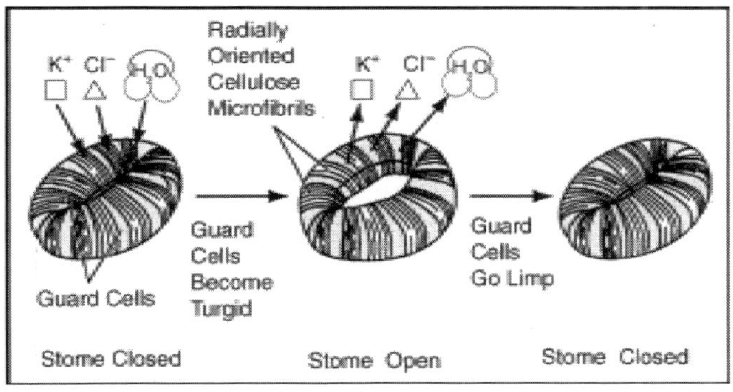

Together, the guard cells and stoma comprise the stomatal apparatus. The inner wall of the guard cell is thicker than the rest of the wall. When a guard cell takes up potassium ions, water moves into the cell, causing the cell to become turgid and swell, opening the stoma. When the potassium leaves the guard cell, the water also leaves, causing plasmolysis of the cells, and a closing of the stoma. Stomata occupy 1% of the leaf surface, but account for 90% of the water lost in transpiration.

TRANSPORTATION AND STORAGE OF NUTRIENTS

Plants make sugar by photosynthesis, usually in their leaves. Some of this sugar is directly used for the metabolism of the plant, some for the synthesis of proteins and lipids, some stored as starch. Other parts of the plant also need energy but are not photosynthetic, such as the roots. Food must therefore be transported in from a source, an action accomplished by the phloem tissue.

PHLOEM, SUGAR, AND TRANSLOCATION

Phloem consists of several types of cells: sieve tube cells (aka sieve elements), companion cells, and the vascular parenchyma. Sieve cells are tubular cells with endwalls known as sieve plates. Most lose their nuclei but remain alive, leaving an empty cell with a functioning plasma membrane.

Companion cells load sugar into the sieve element (sieve elements are connected into sieve tubes). Fluids can move up or down within the phloem, and are translocated from one place to another. Sources are places where sugars are being produced. Sinks are places where sugar is being consumed or stored. Food moves through the phloem by a Pressure-Flow Mechanism. Sugar moves (by an energy-requiring step) from a source (usually leaves) to a sink (usually roots) by osmotic pressure. Translocation of sugar into a sieve element causes water to enter that cell, increasing the pressure of the sugar/ water mix (phloem sap). The pressure causes the sap to flow towards an area of lower pressure, the sink. In the sink, the sugar is removed from the phloem by another energy-requiring step and usually converted into starch or metabolized.

Plants Respond to External Stimuli

One plant response to environmental stimulus involves plant parts moving towards or away from the stimulus, a movement known as atropism. Nastic movements are plant movements independent of the direction of the stimulus.

Alterations in Growth Patterns Generate Tropisms

Charles Darwin and his son Francis studied the familiar reaction of plants growing towards light: phototropism. The Darwins discovered that the tips of the plant curved first, and that the curve extended gradually down the stem. By covering the tips with foil, they prevented the plant from curving. They concluded that some factor was transmitted from the tip of the plant to the lower regions, causing the plant to bend.

We now know, from the 1926 experiments of Frits Went, that auxin, a plant hormone produced in the stem tip (auxins promote cell elongation), moves to the darker side of the plant, causing the cells there to grow larger than corresponding cells on the lighter side of the plant. This produces a curving of the plant stem tip towards the light, a plant movement known as phototropism.

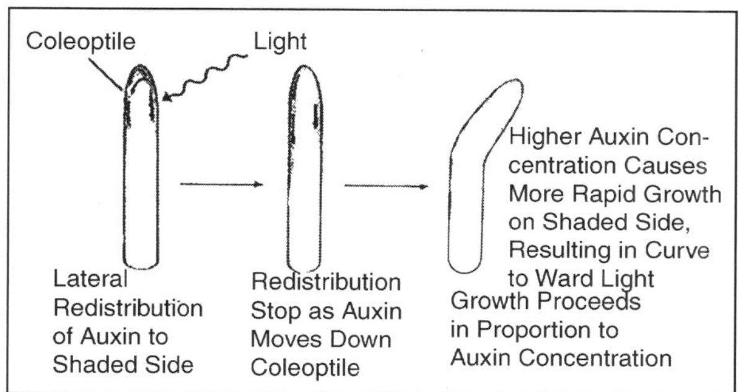

Geotropism is plant response to gravity. Roots of plants show positive geotropism, shoots show negative geotropism. Geotropism was once thought a result of gravity influencing auxin concentration. Several new hypotheses are currently under investigation.

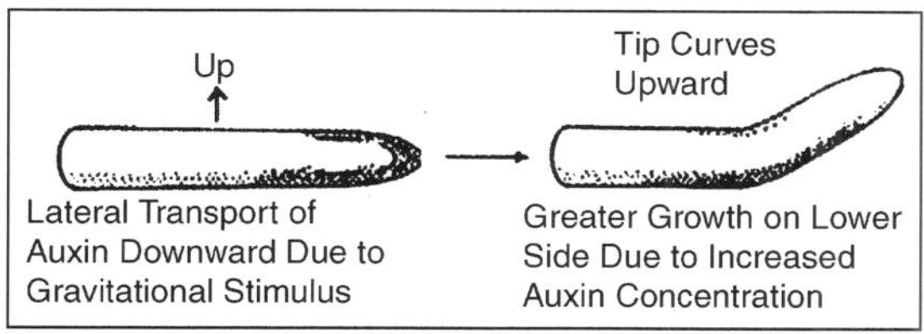

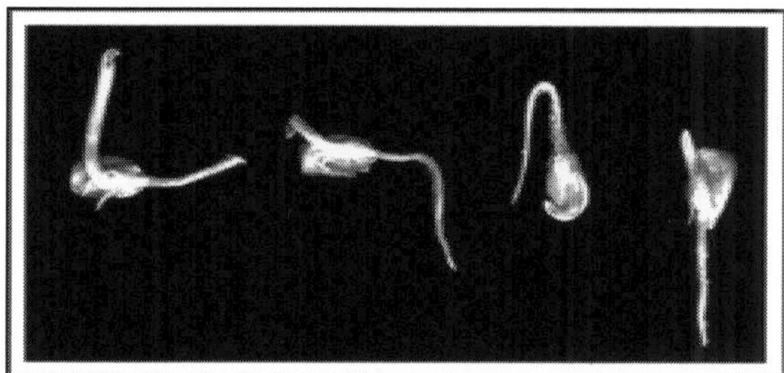

Germination of corn seeds occurs regardless of the seed orientation. Thigmotropism is plant response to contact with a solid object. Tendrils of vines warp around objects, allowing the vine to grow upward.

Curling of a tenbdril around a metal support, an example of thigmotropism. Note the tendril of this passion flower wrapping around the metal rod. Nastic movements, such as nyctinasty, result from several types of stimuli, including light and touch. Legumes turn their leaves in response to day/night conditions. Mimosa, also known as the sensitive plant, has its leaves close up when touched.

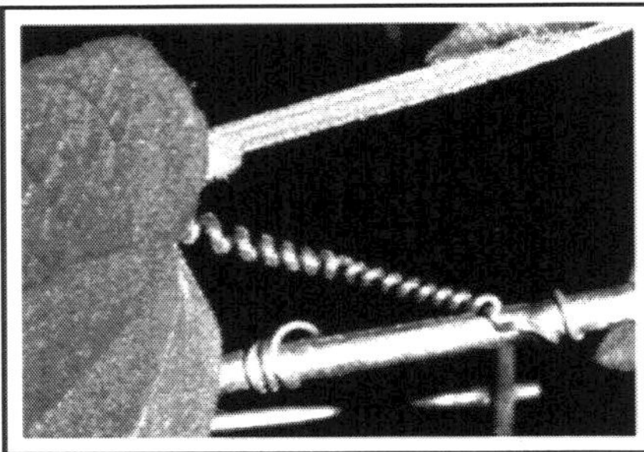

Photoperiodism is the plant response to the relative amounts of light and dark in a 24 hour period, and controls the flowering of many plants. Short-day plants flower during early spring or fall, when the nights are relatively longer and the days are relatively shorter. Long-day plants flower mostly in summer, when the nights are relatively shorter and the days are relatively longer. Day-neutral plants flower without respect for the day length. Phytochrome is a plant pigment in the leaves of plants that detects the day length and generates a response.

PLANT SECONDARY COMPOUNDS

Plants produce primary compounds important in their metabolism. They also produce secondary compounds that serve to attract pollinators, kill parasites, and prevent infectious diseases. Pea plants produce pisatin, a chemical that protects them from most strains of parasitic fungi. Some strains of the fungus (*Fusarium*) contain enzymes that inactivate pisatin, allowing them to infect pea plants.

Some plants produce natural insecticides, such as pyrethrum, a chemical produced by chrysanthemums that is also commercially available to gardeners. Antinutrients are chemical produced when plants are under attack. These compounds inhibit the action of enzymes in the insect's digestive system.

More that 10,000 defensive chemicals have been identified, including caffeine, phenol, tannin, nicotine, and morphine.

Some plant secondary compounds are useful to humans as:

- Pesticides
- Medicines (salicylic acid, the main component in aspirin)
- Stimulants (caffeine, the most widely used psychoactive drug in the world)
- Chewing gum (chicle, a compound from the sapodilla tree in Mexico was used in the first chewing gum).

ESSENTIAL NUTRIENTS IN PLANTS

Of all the essential nutrients, nitrogen is required by plants in the largest quantity and is most frequently the limiting factor in crop productivity.

- In plant tissue, the nitrogen content ranges from 1 and 6 per cent.
- Proper management of nitrogen is important because it is often the most limiting nutrient in crop production and easily lost from the soil system.

NITROGEN FORMS AND FUNCTION

Forms of nitrogen available for plant uptake:

- Ammonium
- Nitrate

Functions of nitrogen in plants:

- Nitrogen is an essential element of all amino acids. Amino acids are the building blocks of proteins.
- Nitrogen is also a component of nucleic acids, which form the DNA of all living things and holds the genetic code.
- Nitrogen is a component of chlorophyll, which is the site of carbohydrate formation (photosynthesis). Chlorophyll is also the substance that gives plants their green colour.
  - Photosynthesis occurs at high rates when there is sufficient nitrogen.
  - A plant receiving sufficient nitrogen will typically exhibit vigourous plant growth. Leaves will also develop a dark green colour.

The Nitrogen Cycle

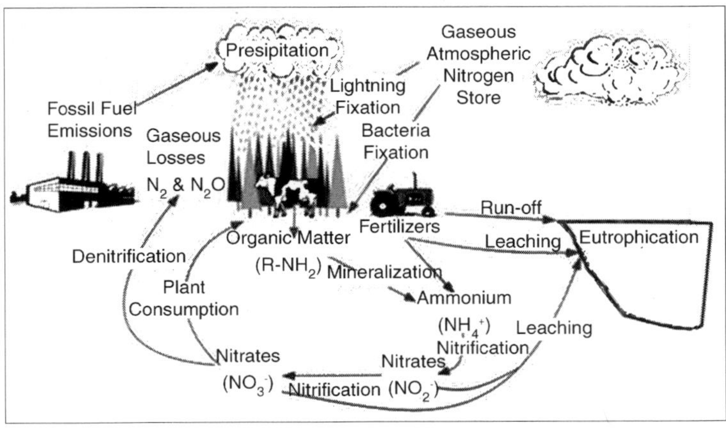

Fig. The Nitrogen Cycle

Gains of Nitrogen to the Soil:

- *Biological and Atmospheric Fixation*: Conversion of atmospheric nitrogen to ammonium which is subsequently available for plant uptake
- Direct additions of commercial and organic fertilizers

Transformations in the Soil:

- Mineralisation: Conversion of organic nitrogen to ammonium
- Nitrification: Conversion of ammonium to nitrate

Losses of Nitrogen from the Soil:

- *Denitrification*: Conversion of nitrate to atmospheric forms of nitrogen
- *Volatilisation*: Loss of gaseous ammonia to the atmosphere
- Run-off
- Leaching
- Consumption by plants and other organisms

Nitrogen is a very dynamic element. It not only exists on Earth in many forms, but also undergoes many transformations in and out of the soil. The sum of these transformations is known as the nitrogen cycle.

Table. Various forms of Nitrogen

Form of Nitrogen	Formula	Availability for plant uptake
Nitrogen gas	N_2	Although 78 per cent of our atmosphere is nitrogen gas, this form of nitrogen must be transformed to usable forms before it is available for plant uptake.
Ammonia	NH_3	Ammonia is a gas. Ammonium can escape from the surface of the soil under certain conditions and is harmful to plants in high quantities. Ammonium is the basic building block of commercial nitrogen fertilizers.
Ammonium	NH_4^+	Soil particles attract and retain ammonium on cation exchange complexes. This form may be directly taken up by plants.
Nitrate	NO_3^-	Nitrate is the second form of nitrogen which is available for plant uptake. In most soils, nitrate is highly mobile. However, in the highly weathered soils, nitrate is stored in soils with 'anion exchange capacity' and becomes less mobile.
Nitrite	NO_2^-	Nitrite is an intermediate product in the conversion of ammonium to nitrate (nitrification). It is usually present in low quantities, but is toxic to plants.

Organic Nitrogen Various compounds Organic nitrogen must be converted to ammonium before it is used by plants. This conversion occurs with time and is known as mineralisation.

Though complex, the nitrogen cycle:

- Helps us to understand the complex relationships that exist between the many forms of nitrogen
- Provides us with insight pertaining to the availability of ammonium and nitrate, which are the only nitrogen forms usable by plants
- To understand the many ways in which N may be lost from the soil

In this section, we will discuss eight major transformations of nitrogen in the soil: nitrogen fixation, mineralisation, immobilisation, nitrification, denitrification, volatilisation, and leaching.

Nitrogen Fixation

Although atmospheric nitrogen gas(N_2) makes up approximately 78 per cent of the air, it cannot be directly used by plants. Instead, atmospheric N_2 only becomes available to plants through three unique processes. The final product of each of these processes is ammonium, which is then available for plant uptake.

The three processes which convert atmospheric nitrogen to ammonium:

- Biological nitrogen fixation
- Chemical nitrogen fixation
- Atmospheric addition

Biological Nitrogen Fixation

Certain soil organisms have the special ability to convert atmospheric nitrogen to ammonium. These organisms include several species of bacteria, actinomycetes, and cyanobacteria.

In the soil, nitrogen fixating organisms can form special relationships with plants, called "symbiotic" associations. Symbiotic is a term that means "living together." Although a symbiotic relationship can be antagonistic, the symbiosis that occurs during biological nitrogen fixation is generally mutual and beneficial.

Legume-Rhizobium Symbiosis

The most abundant symbiotic relationship in nitrogen fixation forms between legumes (*i.e.*, alfalfa, soybeans, etc.) and the *Rhizobia* bacteria species.

- As the roots of legumes grow, *Rhizobium* bacteria infect the root hairs where they begin to multiply.
- As a response to this colonisation, the legume forms nodules, which are structures that form around the *Rhizobia*.
- Within these nodules, *Rhizobia* bacteria are able to continue multiplying and converting the N_2 from the soil air to ammonium.
  - However, the presence of nodules is not a sufficient indicator that nitrogen is being converted to ammonium. Active and effective nodules are generally greater than 2 mm, have pink to red interiors, and concentrate around the tap root.

– On the other hand, non-effective nodules are generally smaller in size with white, green or brown interiors.

The image below shows cross sections of soybean (*Glycine max*) nodules. Nodules in the first row are highly effective at converting N_2 from the atmosphere into ammonium. The nodules in the second row are moderately effective at biological nitrogen fixation, while the bottom row of nodules do not fix any nitrogen at all. Notice the lack of colour in the interior of these nodules, which indicates that they do not have an active 'nitrogenase system'. A 'nitrogenase system' is the bacterial enzyme that is necessary to convert N_2 gas into ammonium through this biological process.

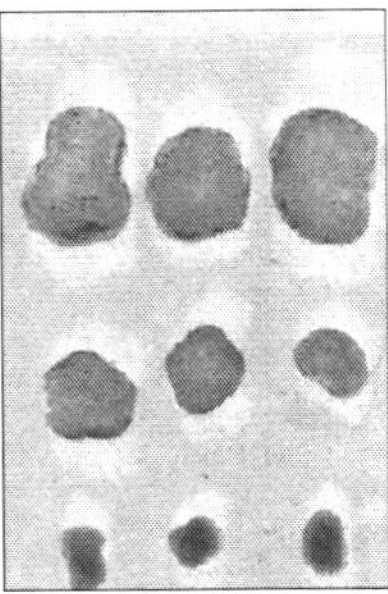

Fig. Relationship between Colour and Effectivity of Nodulation in Soybean.

The next image shows how a well nodulated soybean root system looks when it contains many highly effective *rhizobia* in the soil or is applied as an inoculant.

Fig. Depiction of Effective Nodulation of Soybean by *Rhizobia*.

Specificity

Some *Rhizobium* species are only capable of nodulating a particular legume species and cannot successfully nodulate other legumes.

- For example, the *Rhizobium* that nodulates alfalfa is a different species from the *Rhizobium* that nodulates soybean.

This phenomenon is known as *Rhizobium* specificity. However, not all *Rhizobium* are legume-specific. Thus, some may nodulate a number of different legumes. The figure below gives examples of some common cross-inoculation groups that may assist you in selecting the proper rhizobial inoculant for a particular legume host. The proper combination of *rhizobia* and legume will result in the optimal nodulation and most nitrogen fixation.

From this figure, we see that using soybean *rhizobia* with a soybean plant forms an effective symbiosis, while using a soybean *rhizobia* with a *leucaena* plant does not. However, cowpea rhizobia is capable of nodulating both mungbean and peanut.

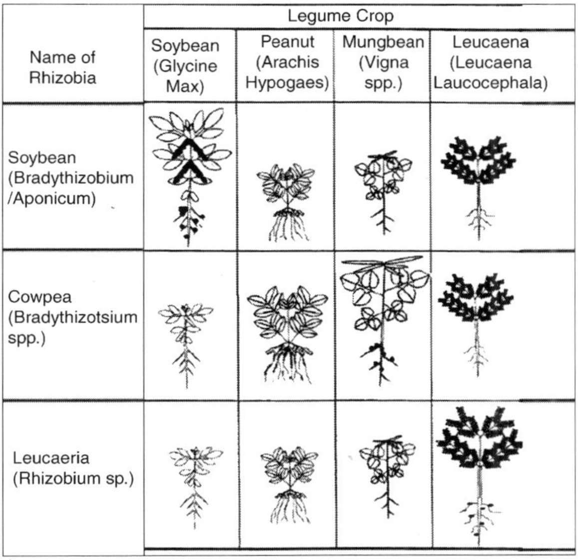

Name of Rhizobia	Legume Crop			
	Soybean (Glycine Max)	Peanut (Arachis Hypogaes)	Mungbean (Vigna spp.)	Leucaena (Leucaena Laucocephala)
Soybean (Bradythizobium /Aponicum)				
Cowpea (Bradythizotsium spp.)				
Leucaeria (Rhizobium sp.)				

Fig. Specificity of Rhizobia for Successful Nodulation of Certain Legumes.

Biological Nitrogen Fixation Management Programme

The occurrence of the symbiotic relationship is heavily dependent upon a variety of soil conditions. If your programme incorporates nitrogen fixation, the following considerations can determine your success.

- First and foremost, the *Rhizobium* must be compatible with the legume. If your crop is *Rhizobium* specific, you must use the correct *Rhizobium* species.
- If your inoculum (which contains the *Rhizobium* bacteria) is applied to seeds, the procedures must be properly followed.
- Nitrogen fixation takes place when total soil nitrogen is insufficient. When sufficiently present, the plant will instead rely on the nitrogen available from the soil.
- *Rhizobia* are sensitive to any growth factor that limits root

development. Such conditions as aluminum and manganese toxicities will limit inoculation.

- *Rhizobia* are influenced by mineral nutrient imbalances.
  - Low levels of calcium, phosphate, molybdenum under acidic conditions will limit nitrogen fixation.
  - Under alkaline conditions, phosphate, cobalt, boron, iron, and copper levels become a concern.
- Any growth factor (such as light, water temperature stresses or soil compaction) and any management factor (such as nutrient management, salinity) that detrimentally affects growth of the legume will detrimentally impact nitrogen fixation.

Table. A Summary of Biological Nitrogen Fixation Measurements by Different Legumes.

Crop Total N (%)	**Location N_2 Fixed**	**Crop N**		**N_2 Fixation (kg N ba^{-1})**
Soybean	Brazil	112-206	85-154	70-80
	Hawaii	120-295	117-237	80-97
Groundnut	Australia	171-248	37-131	22-53
	India	126-165	109-152	86-92
Common bean	Brazil	18-71	3-32	16-71
	Kenya	128-183	17-57	16-32
Cowpea	Indonesia	25-69	9-51	12-33

Inoculation of Legumes with *Rhizobia* Increase Yield and Biological Nitrogen Fixation

This question is commonly posed by farmers who are deciding when to apply an inoculant to their crops. Since, inoculation is relatively inexpensive (less than $5.00/acre), growers should error on the side of caution and inoculate their legume crops unless they have good evidence that inoculation is not needed.

The figure below is a conceptual model which integrates all the factors controlling biological nitrogen fixation.

Additionally, it explains when the inoculation of legumes with *rhizobia* will result in an increase in plant growth and biological nitrogen fixation activity.

The model is based on the fact that if the plant's need for nitrogen is greater than the nitrogen that is supplied by both the existing soil nitrogen and the *rhizobia* already present in the soil, the inoculation of supplementary effective *rhizobia* will result in increased yield and biological nitrogen fixation.

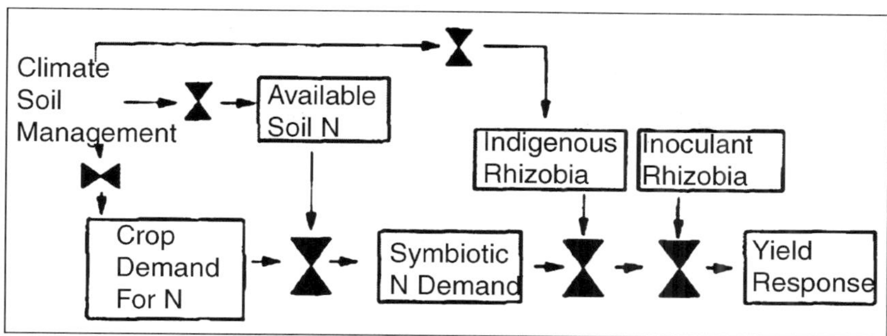

Fig. The Various Factors that Control Nitrogen Fixation.

Amount of Nitrogen Fixed by Legume

When nitrogen is converted to ammonium during biological nitrogen fixation, ammonium becomes available to the legume and the microorganism that fixes it.

Typically, the bacteria can fix anywhere between 20 and 80 per cent of the total legume N.

- Perennial legumes may fix 100 to 200 lb/a/yr
- Annual legumes fix 50 to 100 lb/a/yr.

Small amounts of ammonium can also be released by roots of the legume into the rhizosphere, or the surrounding soil.

Availability of Nitrogen in Subsequent Cropping Systems

Research shows that yields of non-legume crops can increase when following a legume rotation. It is believed that the legume rotation increases the N content of soil, thus making it an effective nutrient management strategy.

However, when the legume is incorporated into the soil, the major benefit of the legume rotation lasts only during the first year following the legume rotation.

Other Symbiotic Relationships

Legumes and *Rhizobia* are not the only species that can establish a mutual symbiotic relationship needed for nitrogen fixation to oocur.

- In wetland rice production, a symbiotic relationship may form between *Anabaena azolla (a blue green algae)* and the Azolla fern. As a result, wetland plants can benefit by incorporations of Azolla as a green manure.
- Certain tree species (i.e Causarina sp.) can form symbiotic relationships with certain species of Actinomycetes and the Frankia bacteria. Though this relationship has lesser agricultural importance, it may gain significance in forestry, or wood production.

Free-Living Nitrogen Fixation

"Free-living" nitrogen fixating organisms are also capable of nitrogen fixation, but are not associated with any plant species.

- Examples of these organisms are azotobacteria, azolospirillum, and clostridium. However, free-living species do not contribute largely to agricultural production

Chemical Nitrogen Fixation

Since, the 1950s, ammonium-based fertilizers have been manufactured using the Haber-Bosch technique. In this catalytic process, N_2 reacts with hydrogen under 1,200 degrees Celsius and 500 atm. Since, the production of chemical fertilizers requires large inputs of fossil fuel, chemical fertilizers can be relatively expensive. The impact of the Haber-Bosch technology on agriculture has been very dramatic. The Haber-Bosch technology enables high-analysis ammonium fertilizers to be produced quickly. As a result, the reliance on biological N fixation and manures as N sources has declined.

Atmospheric Nitrogen Additions

Nitrogen is deposited onto the earth's surface by:

- Rain
  - In the form of ammonium, nitrate, nitrite
- Finely divided organic N swept along the earth's surface
- Lightning
  - Responsible for approximately 10-20 per cent of soil nitrate (114, Fertilizers)
- Industrial wastes
  - As compared to the major industrial regions of the Mainland, industrial wastes do not significantly contribute to atmospheric N.

This table presents estimates of the different sources of atmospherically fixed nitrogen that was deposited onto the earth in the latter half of the twentieth century. Biological sources account for around 20 per cent of total nitrogen deposition.

Biological N Year^{-1}	Million Tonnes	Non-biological N Year^{-1}	Million Tonnes
Agriculture		Industrial	70
Legumes	35	Combustion	61-251
Rice	4	Atmospheric	131-321
deposition			
Grassland	45		
Other crops	5		
Forest	40		
Other	10		
Total	139	Total	262-642

Nitrogen Mineralisation in Soils

When absorbed by plants, ammonium and nitrate are incorporated into plant cells as organic, or living, forms of nitrogen. When plants die, microorganisms break down, or decompose, dead plant cells. During the decomposition about plant matter, organic nitrogen is once again converted to inorganic ammonium and released into the soil.

The process that converts organic N to ammonium is called *mineralisation* and plays a significant role in the management of nitrogen.

Conditions Affecting N Mineralisation

The amount of ammonium that is released to the soil through mineralisation depends on several factors:

- *Quantity of Organic Nitrogen*: The amount of organic nitrogen originally present in the organic matter determines the amount of N that can ultimately be mineralised.
- *Temperature*: The optimal range for mineralisation to occur is between 77-95 degrees Fahrenheit.
- *Oxygen*: Microorganisms need oxygen and since, microorganisms mediate mineralisation, sufficient oxygen must be available in the soil.
- *Moisture content*: Ideally, water should fill 15 – 70 per cent of pore space for maximum mineralisation. This roughly corresponds to field capacity.
- *Ratio of carbon to nitrogen (C:N)*: The C:N ratio is a term used to describe the relative amount of total carbon in comparison the amount of total nitrogen present in the soil and/or organic matter.
- This ratio is very important in determining the rate of mineralisation that should occur for a given type of organic matter.

Since, the microorganisms living in the soil need both carbon and nitrogen, net mineralisation occurs when C:N ratio is less than 20:1. This means for every two parts of carbon, there should be 1 part nitrogen for net mineralisation. If you are applying organic amendments to your soil, it is important to become familiar with the C:N ratio to ensure N availability.

After mineralisation of N occurs, ammonium can be:

- Taken up by the plants
- Consumed by other organisms
- Nitrified
- Volatilised

Immobilisation

Immobilisation is process that converts inorganic nitrogen to organic nitrogen. It is the reverse reaction of mineralisation.

Immobilisation occurs when decomposing organic matter contains low amounts of nitrogen. Thus, immobilisation occurs if the source of organic matter has a high C:N ratio. Microorganisms, who also need nitrogen to live, scavenge the soil for nitrogen when plant residues contain inadequate amounts of nitrogen. As inorganic ammonium and nitrate are incorporated into the cells of living microorganisms, the total N levels in the soil are reduced. Immobilisation can ultimately result in nitrogen deficiencies.

When nitrogen is immobilised in the soil, there may be little nitrogen available for crop growth. As a result, plants can suffer from nitrogen deficiency and develop a yellow colouration. This is the reason why organic materials with a high C:N ratios, such as grass clippings and grain stover, are usually composted before they are incorporated into the soil or planting. This allows time for soil microorganisms to decompose the materials and begin to release nitrogen and other nutrients back into the soil

Mineralisation or Immobilisation

The processes of mineralisation and immobilisation are constantly occurring simultaneously. As organic matter decomposes, inorganic nitrogen will be released into the soil. As both plants and microorganisms grow, they utilise the nitrogen in the soil. Once plants and microorganisms die, they decompose and release inorganic nitrogen to the soil through mineralisation.

Even though mineralisation and immobilisation are both occurring, we can determine which process, mineralisation or immobilisation, predominates.

- When mineralisation occurs at a greater rate, we say that there is net mineralisation.
- Likewise, when immobilisation occurs to a greater extent, there is net immobilisation.

Whether there is Net Mineralisation or Net Immobilisation

The answer is the C:N ratio of the decomposable organic matter.

The C:N ratio is a characteristic of all organic matter, which includes:

- Crop residues
- Soil organic matter (including humus)
- Soil microorganisms (remember microorganism have both carbon and nitrogen)

Rule of Thumb

- When the C:N ratio of decomposing organic residues is between 20:1 and 30:1, mineralisation and immobilisation occur at fairly equal rates.
- Net mineralisation occurs at C:N ratios less than 20:1.
- Net immobilisation occurs at C:N ratios greater than above 30:1.

- Most well decomposed organic matter in soils have a C:N ration near 10:1

Management of Organic Residues

If your programme involves the addition of organic residues, it is important to know its C:N ratio. This knowledge allows you to predict whether net mineralisation or net immobilisation will occur. If the residue has a wide C:N ratio range, it may be necessary to apply additional amounts of nitrogen to your soil or choose a residue with a narrower range.

Nitrification

In most aerobic soils under optimal soil conditions, ammonium is rapidly converted to nitrate by soil bacteria through a process known as nitrification.

- Nitrification involves two steps:
  - First, ammonium is converted to nitrite
  - Then, nitrite is converted to nitrate.

As you can see from the outline of steps above, the intermediate product of nitrification is nitrite. If conditions are unfavourable to undergo the second step of nitrification, nitrite can leach into the ground water and pose as a health risk. The process of nitrification produces hydrogen ions. When large quantities of ammonium-containing fertilizers are applied to soil over time, this process can acidify the soil. See figure below for a simplified presentation of the nitrification process.

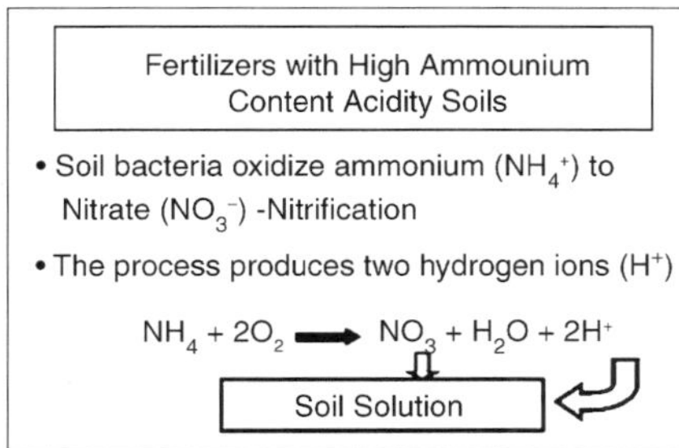

Fig. Basic Process that Causes Soil Acidity by Ammonium Fertilizers.

Factors Affecting Nitrification

There are many factors that affect nitrification. Since, nitrification is mediated by microorganisms, environmental factors that affect biological life will also influence nitrification. In general, the optimal conditions for most plant growth are also the optimal conditions for nitrification:

- *Presence of ammonium in the soil*: In order for nitrification to occur, there must be a source of ammonium in the soil. Sources include mineralised ammonium or additions of ammonium-containing synthetic fertilizers

- *Presence of microorganisms*: Microorganisms that carry out nitrification must be present in the soil.
- *Soil pH*: The optimal pH for nitrification is 8.5, but it may occur over a fairly wide pH range. However, acidity (less than 5.5) has a detrimental effect on the nitrifying bacteria, thus reducing nitrification.
- *Soil moisture*: Nitrification is optimal at the field capacity of the soil. Nitrification is reduced at moisture levels greater and below field capacity.
- Field capacity is the amount of water that remains in the soil after free drainage in a saturated soil ceases.
- Field capacity is also the optimal soil moisture for most plant growth.
- *Soil aeration*: Nitrification requires oxygen. Any management factor that improves soil aeration, such as adding organic matter, will help optimise nitrification.
- *Soil temperature*: Nitrifying bacteria are sensitive to temperature. The optimal temperature range for nitrification is between 77 and 95 degrees Fahrenheit. However, nitrification can occur between 41 and 95 degrees Fahrenheit.

Environmental Considerations

Nitrate is generally a very mobile in most soils. Excessive amounts of nitrate that are not taken up by plants is subject to leaching. Nitrate leaching can have an adverse effect on the environment.

- Nitrate, which moves through the soil profile during periods of rain, pollutes ground water reserves.
- Surface run-off of nitrate is a source of eutrophication, or algal blooms, in lakes and estuaries.

PLANT EFFECTS ON SOIL PHYSICAL CONDITION

While soil structure and related properties certainly affect plant growth, it is also recognized that vegetation can modify physical conditions of soil. Species reported to be detrimental to soil structure are maize, soybeans (Glycine max L.), and cereal species in general. Plant species identified as beneficial to soil aggregation include ryegrass, bromegrass (Bromus ssp.), alfalfa (Medicago sativa L.) and clover (Trifolium ssp.). Stone and Buttery found that nine forages differed in their ability to improve structure. Warm-season C4 prairie graminoids appear to promote aggregation better than cool-season C3 grasses, possibly due to differences in total biomass production, length and timing of growth, physiological differences affecting root exudates, root morphology, and/or microbial population (especially mycorrhizal infection).

Moreover, land uses associated with continuous plant cover of soil and lack of soil manipulation provide greater opportunity for the development of stable soil structure. Use of land as pasture has been found to increase aggregation of previously tilled soil.

MECHANICAL BINDING

One documented mechanism of roots' influence on soil structure is mechanical binding of soil particles by fine roots and microbial hyphae. Networks of dead and especially living roots resist compactive loads and shear stress; fungal hyphae work similarly within soil aggregates. Rooting patterns (*i.e.*, total root length, distribution, root length density, branching frequency, and root hairs) are thought to be important in explaining differences in soil stabilisation by different species. Alfalfa and clover improve soil structure primarily through enhancement of infiltration rate, hydraulic conductivity, or soil water retention/drainage, which might be expected from the channel development under these tap-rooted plants. However, mycorrhizal hyphae associated with the roots of white clover (T. repens L.) have been related to increased stability of aggregates.

Organic Matter

Total soil organic matter content is the soil property most closely associated with soil structure stability. Soil organic matter accumulates over the long term to a steady-state level, which is determined by the amount of biological contributions over time, soil water content and temperature (regulating decomposition), and other factors such as texture. As primary producers in terrestrial ecosystems, plants ultimately can be credited with nearly all of the organic matter added to soil. Direct contributions occur from seasonal shedding of leaves and roots and root exudates as well as the whole plant upon death. The organic compounds added and the microbial activity and products that result greatly enhance soil structure and improve structural stability.

Qualitative differences in soil organic matter and the mechanisms of stabilisation involved may account for unexplained differences in aggregate stability associated with plant species. Often much of the short-term increase in organic carbon has been found in the sand-size fraction, which includes fragments of plant tissue. However, the effects of organic matter on the formation, maintenance, or degradation of soil structure are more directly related to metabolic compounds and decomposition products that interact with soil particles, especially clays, on the molecular level. Hydrophobic coatings (presumably waxes from plant roots or associated microorganisms) can cause water repellency of sand-textured aggregates (94) and lead to the development of localised dry-spot formation in turf. Water repellency can reduce the rate of clay-aggregate wetting and therefore increase aggregate stability and contribute to development of preferential flow paths of infiltrating water. Grassland soil aggregates exhibit greater potential water repellency than aggregates of arable (maize) land. The grassland vs. cultivated comparison, then, may reflect breakage of hydrophobic coatings (exposing uncoated soil) and/or microbial oxidation of the coatings in the cultivated land. Another possibility is a species difference in deposition of hydrophobic coatings on soil particles.

Strong correlations occur between soil carbohydrate content, or some fraction thereof, and soil structural indices. Periodate-sensitive materials (polysaccharide and/or polyuronides) have been shown to be stabilizing agents for aggregates in many cases as well as pyrophosphatesensitive materials probably bound to minerals by polyvalent cations. Carbohydrates have been found to constitute 8 to 16 per cent of soil organic matter in some virgin soils, the amount generally increasing with clay content. Soil in agricultural land use generally contains greater carbohydrate content than fallow, and species variability is evident. Some of the soil carbohydrate is contributed by structural components of plants (*e.g.*, cellulose), but soil also gains carbohydrates in the form of soluble compounds released from plant roots.

Root Exudates

Roots of many species have been shown to exude metabolic compounds of varying composition and quantity. Barber and Gunn's minimum estimate of exudates released was 9 per cent of the dry matter of the root increment grown. Rovira estimated that 0.1 to 0.4 per cent of carbon assimilated by photosynthesis was released to the soil. Carbohydrates are major constituents of root exudates; for example, Moody *et al.* found that water-soluble components of root ''slime'' were mostly (> 95 per cent, w/w) carbohydrate. While some root exudation is considered passive loss of photosynthates by the plant, it is clear that some exuded compounds are produced by the plant specifically to aid root function, induced by environmental conditions.

Therefore, root exudation is influenced by both genetic and environmental factors, including plant species, plant age, temperature, light, plant nutrition, soil moisture, root damage, and foliar applications. Shoot harvest has been shown to affect the amount of alfalfa root exudation. An additional factor that is especially relevant here is mechanical stress, which can increase root exudation of amino acids and carbohydrates. Greater development of secretory organelles in root cap cells under conditions of mechanical impedance presumably ameliorates adverse effects by production of mucilage, wetting the soil. Organic compounds released by roots serve as rich sources of carbon and energy for many soil microorganisms living in the rhizosphere. The term rhizosphere refers to the zone of high microbial activity, compared to bulk soil, that surrounds roots. Further classification of the rhizosphere is defined by root products and abundance of microorganisms.

The microbial community that develops in the rhizosphere may be characteristic of the plant species. These microbial communities produce their own "exudates," which, in turn, can be influential in particle binding. Relative contribution of microorganism- vs. plantderived carbohydrates to soil has been a subject of study. Direct effects of root exudates on aggregate stability has shown variability; for example, extracted bromegrass root exudates increased wet aggregate stability and decreased dispersible clay compared to maize exudates. Quantitative/qualitative differences in root exudates are likely responsible for species' effects on soil aggregation. Soil aggregate stabilisation by ryegrass was maximized by shoot harvest, perhaps by increasing root exudates. Maize presents a special case in terms of root contributions to soil and subsequent soil structure effects. Cheshire and Mundie claimed that as little as 0.5 per cent of 14C fixed by maize during 36 days was released as watersoluble substances; however, maize roots are characterized by abundant mucilage or mucigel.

Low-molecular-weight components exuded from root cap cells polymerise extramurally and hydrate to form mucilage, which is likely to remain associated with mineral surfaces after dispersion. An SEM study of maize roots indicated a mucilage layer nearly covering the entire length, with soil aggregates embedded in the mucilage at the root cap and on root hairs. However, soil carbohydrate content and soil structure are usually not improved by maize culture. Evidence that maize-root compounds are more readily decomposed may reflect the distinct composition of maize-root exudates and explain lack of soil stabilisation. One theory regarding the detrimental effect of maize on soil aggregates suggested that chelating agents released by roots disrupt organic matter–mineral particle bonds. Pojasok and Kay disputed this, finding that the relative degree of aggregate stabilisation by bromegrass or maize correlated with the amount of divalent cations released from soil in response to root exudates. Tillage of soil that is common for maize production complicates field studies.

Soil Drying

Soil drying by root extraction of water has been indicated as a stabilizing force in several studies. Reid and Goss speculated that drying promoted adsorption of organic materials onto mineral surfaces. Wetting and drying of molded soil increases the stability of the molded form, which is analogous to the drying of puddled soil in the field. A logical explanation of this phenomenon is optimisation of particle position and orientation to maximize bonding forces as water films thin and particles are slowly drawn together. Naturally formed aggregates, however, appear to be more strongly influenced by biological activity than by drying forces.

Time

Time is an important factor in the stabilisation of soil aggregates by plant influences. Only 8 weeks were necessary before increases in the aggregate stability with ryegrass could be measured, but the effect leveled off at 16 weeks. In another study, alfalfa and ryegrass caused linear increases in soil aggregation from 1 to 4 years; but a degraded soil, such as has been under continuous row cropping, did not recover aggregation levels equal to a virgin site even 30 years after conversion to meadow.

Reduction of infiltration into a compacted subsurface layer under turf could still be seen 12 years after the compacted layer was developed despite a trend towards faster infiltration with age. Biologically and physically significant recovery of aggregation occurred after 5 to 10 growing seasons in cultivated soil restored to tallgrass prairie at a site in eastern Illinois. Rapid recovery in this case was aided by high degrees of initial aggregation, very high production of biomass (1 to 1.5 kg m^{-2}), lack of soil manipulation, and vegetation type with associated microorganisms.

Separating the effect of time without disturbance from the effect of vegetation type was not entirely possible, but such effort indicated that the time factor might predominate. A conceptual model of soil-structure dynamics proposed by Gibbs and Reid emphasised macropore (> 100 μm) dynamics; and the direct root effects on macroporosity included in the model were creation and blockage of macropores and roots' decomposition to organic matter.

Activities of living roots were cited as indirect influences on organic-matter humification, macropore stability, shrinkage (due to water uptake), and planar micropore expansion by mechanical pressure.

Roots Plugging Pores

Roots potentially can be a contributing factor in the reduction of soil aeration. Substantial external force is applied by plant roots as they grow; the pressure of expanding roots may approach 0.9 MPa, causing a zone of compaction around the root and reorienting soil particles tangential to the root. Bruand *et al.* also demonstrated quantitatively the reduction of porosity in the radial vicinity of roots. Backscattered electron scanning images of soil porosity around maize roots revealed that porosity was 22 to 24 per cent less within the soil surrounding the root than in the bulk soil, and the bulk density increased up to 1.80 Mg m"3 at the root–soil interface, whereas density of the bulk soil was 1.54 Mg m^{-3}.

Micropore collapse was suggested to be induced through water extraction by the root as well as through root expansion. Dexter developed a simplified model for soil compression around roots, using the assumptions that the root volume is accommodated by loss of porosity in the surrounding soil, that there is a minimum soil porosity below which soil will not be compressed, and that density decreases exponentially with distance from the roots' surface, the exponent being a constant multiple of the root diameter. Asady and Smucker found that roots occupying more than 5 per cent of the air-filled porosity of a soil resulted in O_2 consumption rates that exceeded the supply rate and increased CO_2 concentrations, because diffusion was decreased by the presence of excessive quantities of O_2 -consuming root sinks.

Thus, soil regions below and adjacent to areas with the greatest accumulation of roots can be prone to reduced aeration, particularly if the soil is also compacted. Studies of turf and forage grasses have reported very high rootlength densities compared to those reported by Asady and Smucker and others working with field crops. Thus, it is appears plausible that plant species with very high root-length density could develop soil oxygen stress due to extensive rooting that plugs air-filled pores and compacts the rhizosphere soil. This condition may be exacerbated under high soil temperatures, when the demand for oxygen by plant roots and soil microbes is greatest.

7

Soil Management

Most soils cannot provide one or more plant essential nutrients in sufficient available form for modern crop production. Soil samples are frequently tested to determine the quantities of nutrients and other amendments which should be applied. Soil testing normally involves extraction or reaction of a sample with a specific chemical solution(s) which removes essential elements in amounts related to those required for plant growth. Soil testing is recommended to prevent both under and over fertilization of crops, thereby providing economic crop production in an environmentally effective manner.

Successful farming does not merely depend upon the knowledge of physical, chemical and biological properties of the soil. It concerns both soil and soil management. The most important consideration is the correct application of the relationship between the soil and the crops to be grown. Although the problems of soil management vary according to the soils and their situations, the climatic conditions and the crops to be grown, there are fundamental factors that govern the choice of soil management practices.

Good soil tilth is the first feature of good soil management. It means a suitable physical condition of the soil and implies, in addition, a satisfactory regulation of soil moisture and air. The maintenance of soil organic matter which encourages granulation is an important consideration of good tilth. Tillage operations and timings should be so adjusted as to cause the minimum destruction of soil aggregates. Good tilth minimizes erosion hazards.

The choice and sequence of adaptable crops or crop rotation are very important. These are related to climate, rainfall, its distribution, and the characteristics of the soil profile, like drainage and soil moisture. A proper sequence of crops and varieties greatly influences the soil physically, chemically and biologically. It is better to evolve cropping patterns according to land capacity classification, but where such information is not available, the recommendations of the local agricultural agencies may be followed. The cropping patterns and management principles should be based on the principles of soil and water conservation and efficient moisture utilization. In irrigated areas special management principles are important to avoid salinity, alkalinity and water-logging. The capacity of the soil should be improved and maintained by adequate organic manures and plant nutrients through fertilizers and including legumes in the rotation. Similarly, the provision of irrigation facility in the semi-arid and arid areas, different remedial measures against excessive salinity and alkalinity, or acidity in humid areas, specific soil amendments to correct plant-nutrient imbalances and the application of trace elements are some of the measures needed in different and special circumstances.

Economic plant-protection measures should form part of the management practices in the cropping system. This is achieved by changing the cultural practices and using fugicides, insecticides etc. The economics of selected management practices is of vital importance. Unless economically profitable the recommended practices are not likely to be adopted. Hence, a package of practices for the integrated land use plan, including all the above points and yet economically profitable is necessary. This is now available for different areas in the country through the All-India co-ordinated schemes, operational research projects and other schemes at centre and state level.

Before suggesting changes in soil management for a holding or a group of holdings, all the available information on different items mentioned above should be collected from field surveys, farms of research stations, farmers and extension workers and analysed to arrive at a rational package of practices for evolving suitable cropping systems, such a package of practices should be developed to cater for the needs of all kinds of farmers-large, small and marginal.

Lime

The main purpose of liming is to raise soil pH and supply calcium and sometimes magnesium for plant growth. Other benefits from liming acid soils include increased biotic activity, enhanced mineralisation of nutrients from soil organic matter, improved soil structure, decreased potential for aluminum toxicity, and increased availability of other nutrients, especially phosphorus.

Fertilization

A good fertilization programme is based on soil testing, especially over several years, so that changes in nutrient availability and other chemical properties can be determined over time. Nutrients can be added in either organic or inorganic forms. Remember that plants utilise only specific ionic forms of the essential nutrients. These specific ionic forms can be provided directly through inorganic fertilizers or indirectly through the decomposition of manures, composts, and other organic amendments. Thus, plant uptake of nutrients should be similar whether initially added through inorganic or organic sources. Additional benefits which may accrue from application of organic materials include improved soil structure and increased water holding and cation exchange capacities. The degree to which these additional benefits occur will be determined by the quantity and quality of organic material applied and its decomposition rate. Municipal sewage sludge, food processing wastes, and other similar organic wastes are commonly added to soils. Besides the obvious need for disposal, these materials can provide nutrients for plant growth and improve soil properties. Sewage sludges can contain heavy metals, however, especially in industrialised areas. Land application of sewage sludge is usually regulated by state and federal guidelines. When properly utilised, the above materials can be suitable sources of plant nutrients.

Irrigation

Irrigation is practiced to at least a limited extent on each of the earth's continents. Irrigation is a method of at least partially overcoming problems in natural precipitation patterns. These problems may result from an overall lack of precipitation or poor seasonal distribution. Irrigation is normally practiced not only to increase yield, but to provide yield stability. Yields may be increased many fold by irrigation depending on climate and the crop produced. Large scale irrigation projects may also be one facet of an overall programme to provide hydroelectric power, flood control, municipal water supplies, and recreation.

Water sources for irrigation include lakes, streams, and groundwater. Numerous water delivery methods are used including furrow, flood, sprinkler, and drip irrigation. Each system is different in efficiency and cost. Both these factors are normally considered before a producer decides on the type of system to be used. Water districts and governmental agencies may also mandate the type of system used based on water quantity, quality, and environmental concerns.

Although development of irrigation capabilities almost immediately increases crop production, long-term effects must also be addressed. Irrigation water must be of reasonable quality (sufficiently low in dissolved salts and of proper ionic composition) to be used for an extended period of years. Many irrigation systems from ancient times through today have failed because of increasing soil salinity over time. Water applied through irrigation is partially transpired by plants or evaporated from the soil. Salts added in the water remain in the soil. Over time, salts can accumulate so as to decrease yields or prevent further crop production.

Providing adequate drainage is also imperative for irrigated lands. Proper drainage not only decreases salt accumulation, but prevents shallow water tables which may restrict plant rooting depth and contribute to upward salt migration during periods of high evaporative demand.

Irrigation has been extremely important throughout the world in providing a stable and abundant food supply, but proper planning and expertise are necessary to sustain economic and environmentally sound irrigated crop production.

Tillage Systems and Conservation Practices

Tillage systems vary widely from conventional, or "clean", tillage where all crop residues are incorporated into the soil with little residue remaining on the soil surface to no-tillage where crop residues are not disturbed and remain on the soil surface from one crop to the next. Conservation tillage denotes any tillage system in which at least 30 per cent of the soil surface is covered by residue after planting. Tillage is used to destroy weeds, incorporate residues from a previous crop, open the soil surface for increased infiltration and aeration, shatter compacted zones which might restrict root growth, and prepare seedbeds for optimal seed germination.

The principal advantage of conservation tillage is protection of surface soil from water and wind erosion and is frequently used to decrease soil loss on erodible lands. Secondary advantages include increased water storage and organic matter contents. Erosion not only decreases the productivity of soils, but additional economic loss is associated with siltation of reservoirs, dredging of waterways, and removal of particulates to meet drinking water standards. As plant residue cover of the soil surface nears 100 per cent, soil erosion decreases to near zero. Erosional loss can be reduced by more than 80 per cent with a residue cover of 50 per cent. Even a modest residue cover of 10 per cent can reduce erosion by about 30 per cent.

Other alternatives are available for erosion control. Construction of terraces, or berms, which slow and alter the movement of water across slopes has been used for centuries to decrease water erosion. Strip cropping, the planting of alternating strips of crops of different heights or seasonal maturities across a landscape, and contour farming, the production of crops across instead of with slopes, have also historically been used to reduce the erosional forces of wind and water.

Rotation, or planting different crop species, is normally beneficial compared to monoculture, or continual production of the same crop species. Rotation often decreases problems from weeds, disease and insects and may improve soil chemical, physical, and microbiological properties, and crop yield.

Farmers must make many decisions each year. These decisions must result in both economic and environmentally prudent crop production if food production is to be sustainable over the long term.

SOILS AND SOCIETAL ISSUES

An escalating world population will certainly result in increased demand for food, fibre, water, and other natural resources. Only one billion persons are estimated to have lived throughout all history up to the year 1850. Contrast this statistic with an estimated world population of seven billion by the year 2000 and the potential problem becomes more sharply focused. World population is currently doubling about every thirty-five years. Will we realistically be able to feed and clothe this burgeoning mass by the year 2025? By 2050? As one of earth's most vital and fragile natural resources, soils come under ever increasing pressure. Quite simply, there is less soil per person with each passing day, making increased productivity per unit of land area a requirement. At the same time there are issues of appropriate land uses, sustainability, environmental protection and water rights.

The Soil Conservation Service (recently renamed the Natural Resources Conservation Service) was established in the United States in 1935 following catastrophic wind erosion of the plowed prairie lands of the Great Plains. Protection of arable land from the ravages of accelerated erosion has been its primary goal. Beginning in 1985, Farm Bills passed by the U.S. Congress have contained provisions that require use of erosion control measures by producers who participate in governmental price support programmes. For many years farmers and ranchers have been encouraged to make cooperative conservation agreements with the Soil Conservation Service. These agreements entitled them to financial assistance for implementing erosion control practices. In much of the developing world, where population growth has been greatest, organised conservation efforts have been weak and pressure on arable soil resources has been great. This growth has led to a tragic loss of valuable resources partially because resources not well suited for crop production are being used as food requirements increase.

Land use planning, not popular with persons who value individual rights in use of owned property, is becoming more widespread as more demands are placed on soil and water resources. Further, concern that land retain its productivity over time is increasing. "Wearing out the soils" on a farm and moving on to a virgin territory is no longer the option that it was well into the present century.

An additional land use issue which will take on increasing global importance is urban development. Many of the world's large cities developed where they did because of the stable agricultural production from productive soils in the area. As cities expand, formerly productive agricultural land is covered by highways, housing subdivisions, shopping malls, and other commercial developments. Once appropriated for such uses, these lands cannot economically be returned to agricultural production. Responsible land use planning is expected to become increasingly important in the future.

Organic gardening and farming have attracted much attention in recent years. As emphasised, the organic component in soils has essential roles in determining the soil's physical, chemical and biological behaviour.

Increasingly, decisions that impact use and management of soil resources arise from discussion of problems such as sustainable use of finite resources, water quality, sedimentation in lakes, food quality, endangered species, preservation of wildlife habitats, and cultural practices. Every citizen, through the right to vote, can influence laws that determine how soil and water resources are used and managed. All citizens share the responsibility for ensuring that all natural resources are treated with respect and wisely used. Some citizens must devote their careers to developing still better technology for management and conservation that will ensure continued utility of these precious resources by our successors on planet earth.

SOIL ANALYSIS

The nutrient status of soils can be determined by chemical analysis, in which the amount of a particular element in a known weight of soil is measured. However, not all of the element present in the soil may be available to plants, so forms of analysis which attempt to mimic the ability of plants to extract nutrients from the soil are commonly used.

For example, the soil may be extracted with a citric acid solution, and the amount of the nutrient in the solution determined. The best extracting solutions have been found by trials in which the results of analysis using different extracting solutions are compared with plant growth observed in field trials.

The available cationic nutrients are generally determined by extracting a known weight of air-dried soil with an extracting solution, and then determining the amounts of K^+, Mg^{2+}, Ca^{2+}, etc, present using atomic absorption or atomic emission spectrophotometers, which measure the amount of light characteristic of a particular element absorbed or emitted when the solution is sprayed into a flame.

The other important parameter that needs to be monitored is the pH of the soil. Soil becomes progressively more acidic over time as a soil is leached by rainwater saturated with carbon dioxide.

The cations are slowly replaced by hydrogen ions in reverse order to the strength of their binding by the soil, and carried away to the oceans by the percolating water. The singly charged ions are lost first, then the doubly and finally the triply charged ions, and the soil progressively becomes more acid in character. Over geological time, even the aluminium and silica is leached away, and the surface of the land is steadily lowered.

This exposes a fresh layer of subsoil to the weathering process, which releases new supplies of nutrients into the newly forming soil in a process of continuous renewal of the topsoil with its nutrients. The physical movement of soil particles, by gravity, wind or water in the process of *erosion* can greatly speed this process, especially on slopes. For gardeners and farmers the soil acidity is a useful indicator of availability of nutrients. This property is often referred to as *soil pH,* though more strictly it refers to the pH of the soil solution. The "soil pH" is readily measured, and gives useful information about the state of a soil. In *acid soils* (pH < 6), many of the exchangeable sites in the clay minerals and humus are occupied by hydrogen or aluminium ions. Remaining nutrient cations are readily released by the soil, but there may not be much of these nutrients present. In very acidic soils, cations such as Fe^{3+} and Al^{3+} appear in the soil solution, and these are quite toxic to many plants. At the other end of the pH range, in basic soils (pH > 8), which are generally derived from limestone, the cation nutrients generally become less available. Plant species have evolved to grow in soils with a particular pH range, and may not thrive in soils with acidities outside this range.

For examples; rhododendrums grow best in strongly acid soils, most trees, shrubs and grasses in slightly acid soils, and many common vegetable crops grow best in neutral to slightly basic soils. The acidity of a soil can be adjusted to suit the type of plants being grown. Most New Zealand native forest plants have developed to grow under the prevailing acid soil conditions, but the soil pH needs to be raised for optimum growth of most crop plants.

The use of acid fertilizers, such as superphosphate, makes the soil more acidic. Soil is normally made less acidic by using lime (finely ground limestone, $CaCO_3$). For faster reaction, "slaked lime" can be used. This is made by heating limestone to form calcium oxide (quicklime, CaO), and then reacting this with water to form calcium hydroxide, $Ca(OH)_2$.

Lime also adds the necessary nutrient calcium, and improves the texture of acid soils (makes them less "sticky") by replacing the H of the -OH groups between the layers of the clay minerals by Ca^{2+} ions. If the soil is deficient in magnesium, the mineral *dolomite*, $CaMg(CO_3)_2$ can be used instead of lime to raise the pH. n the rare occasions when the soil pH needs to be lowered, *e.g.* to grow rhododendrons, this can be achieved by applying sulfur or ammonium sulphate.

These are oxidized, releasing hydrogen ions and so making the soil more acidic:

$$S(s) + 3O_2(g) + 2H_2O(l) \rightarrow 2SO_4{}^{2-}(aq) + 4H^+(aq)\ NH_4{}^+(aq)$$
$$+ O_2(g) + H_2O(l) \rightarrow NO_3{}^-(aq) + 6H^+(aq)$$

GENERAL ANALYSIS

First look at where the soil is, and where it came from. Was the soil formed in position by weathering of the local substrate? What is the nature of the local substrate (sedimentary or igneous rock, volcanic ash, sand, - -)? Or has it been deposited from elsewhere; by the sea, by a river, by the wind, or by human activity? Or was it formed in peat, which is the largely organic residue of an ancient swamp? Much can be learned about a soil just by looking at it and touching it. Look at the soil profile of a freshly cut bank, or dig a hole about 40 cm deep, and study the soil profile. Note the colour and feel of the soil, and the way these change with depth.

Note the amount of organic material (plant debris) present—high for peaty soils to low for clay subsoils. Soils with much organic material are generally dark coloured, soils with little organic material but high in iron(III) oxide are yellow and those low in iron are grey. Rub a sample between your fingers; is it fine and smooth or coarse and gritty? Does it stick to your fingers, or rub off cleanly? Shake a sample of the soil up with water and let it settle; sand and coarse unweathered mineral particles will settle rapidly, fine clay particles will stay suspended as "muddy" water for a long time, while coarse organic material may even float. Try to find what can be deduced from these properties for your local soil.

Animal Life

Carefully crumble about 100 g of freshly collected soil and look for larger animals, such as earthworms and grass grubs. Spread the soil over a sieve (or piece of mesh curtain). Place the sieve over a piece of paper, cover, and leave in a dark place overnight. Try to identify the livestock which falls onto the paper.

Soil pH

Place about 10 g (a teaspoonful) of soil in about 100 ml of distilled water (or fresh rainwater) in a clean stoppered bottle and shake occasionally for about 5 minutes. Allow the solids to settle then gently press a glass electrode into the sediment layer and use a pH meter to determine the pH of the solution. If you do not have access to a pH meter, dip a non-bleeding pH "stick" into the solution and compare the colour developed with the chart; or filter the solution through fine filter paper (or centrifuge the solution) to produce a colourless solution then add a few drops of universal indicator, and compare the colour with the chart to determine the pH. What types of plants will grow well in soil with this "soil pH"?

Available Phosphate Determination

Phosphate is usually analysed by using a reaction with molybdic acid in the presence of a reducing agent. An intensely blue coloured compound (phosphomolybdic acid) is formed, and the colour intensity is proportional to the amount of phosphate present. The colour is compared with that developed in solutions containing a known amount of phosphate. For analysis of an inhomogeneous material such as soil the collection of a *representative sample* is extremely important.

The phosphate level will vary from place to place, and with depth. You could choose a sample from some marked place (and the results would apply only to that place) or collect samples from typical places in an area, *e.g.* from hill-sides, from the level, near trees, etc, etc, and then thoroughly mix them, to get an average value for the area. The usefulness of the result will depend on how representative of the whole area your sample is. Collect a soil sample (or samples). Allow to air dry for a few days so that reproducible results can be obtained. Roughly grind the soil using a pestle and mortar (thoroughly mix if average samples are being used). Accurately weigh out about 0.5 g of soil and place in a stoppered flask, Add 100 mL of the extractant solution A, and shake (or stir with a magnetic stirrer) for 30 minutes. Filter through a fine filter paper. Pipette 50 mL of the filtrate into a 100 mL volumetric flask, and add about 30 mL of water. Add 8 mL of reagent C and make up to 100 mL. The blue colour is developed over 10 minutes and is stable for 24 hours. Pipette 0, 5, 10, 15, 20 and 25 mL of the standard solution E into labelled 100 mL volumetric flasks and fill to about 80 mL with water. Add 8 mL of reagent C and make up to 100 mL. Compare the intensity of colour of the unknown and standard solutions.

If you have a colourimeter, use a blue filter. If you have a spectrophotometer, the maximum absorption is at 880 nm, though a secondary maximum at 660 nm can be used. If you have no instrument, look sideways through the flasks at white paper and try to interpolate the unknown among the standards (or better, pour each solution into a *marked* test tube, filling them all to the same depth, and look down the tubes at white paper. Estimate the extractable phosphate content of the solution and hence of the soil: ìg P g^{-1} of soil = 4 x ìg P ml^{-1} in solution.

Phosphate Retention by Soil

The ability of soils to bind phosphate gives useful information about the likely effectiveness of addition of phosphate fertilizers. In this test the soil is shaken with a standard phosphate solution, and the amount of phosphate adsorbed is determined by analysing the residual solution. The test is most useful for comparing different soils. Accurately weigh about 5 g of ground air dried soil and place in a stoppered container. Add 25 mL of reagent F (which contains 1 mg P mL^{-1}) and shake or stir for 24 hours. Filter through fine filter paper. Pipette 5 mL of the filtrate into a 500 mL volumetric flask, and make up to the mark.

This solution would contain 10 ìg P mL^{-1} if the soil had not adsorbed any phosphate. Prepare calibration solutions by pipetting 10 and 25 mL of solution F into 50 mL volumetric flasks, and make up to the mark. Pipette 5 mL of each of these solutions, and of solution F, into labelled 500 mL volumetric flasks and make up to the mark. Pipette 5 ml of the analysis solution, and of each of the calibration solutions, into separate labelled 100 ml volumetric flasks, add about 80 ml of water and 8 ml of reagent C to each and make up to the mark. Compare the blue colours after 30 min as for the previous test. The calibration solutions correspond to 80, 50 and 0 per cent phosphate retention, respectively.

SOIL MANAGEMENT PRACTICES

AGRICULTURAL PRACTICES AFFECT SOIL ANIMALS

Agriculture generally decreases the number of individuals in the soil and decreases the diversity of soil animals (number of species) compared to native vegetation. Soil animals require certain soil conditions to grow and survive.

Agricultural practices alter soil conditions, making them harsher than those of native vegetation:

- Soil moisture is decreased and temperature is increased
- Wetting/drying cycles are faster
- There is less organic matter in the soil and often only from one or two plants (not a range)
- Soil may be disturbed by erosion, vehicle traffic and tillage

The harsher environments mean that some of the original species are not able to survive. This decreases the species diversity and may cause different organisms to dominate. Intensive tillage practices decrease the diversity of soil animals, especially the larger soil animals, and can interrupt their life cycles. Crop type also has an effect on soil animals, with the numbers in soil decreasing in order of continuous pasture, wheat/lupin rotation and continuous wheat.

Soil Management Affect the Fungal and Bacterial Populations in Soil

Fungi and bacteria differ in their responses to changes in agricultural management practices. Fungi are usually more sensitive to these changes. The fungal-to-bacterial ratio is therefore an indicator of environmental changes in the soil. When plant residues are applied as mulch, for example, fungi prosper because their hyphae are able to grow into the litter layer. Tilling, however, destroys large amounts of the fungal hyphae. Incorporation of plant residues into the soil also favours the bacterial population because the contact surface between the substrate and bacteria is increased. This response further depends on the soil type. Nevertheless, the dominance of either fungi or bacteria also depends on the quality of the plant residue. Substrate structure, C:N ratio and cellulose content are important characteristics of its quality. Fungi are the predominant cellulose decomposers, even though one group of bacteria, the Actinomyces also contribute significantly to its decomposition. Cellulose has a high carbon content and a corresponding high C:N ratio, making it the ideal food source for fungi. Bacteria, which have a smaller C:N ratio than fungi, need food rich in nitrogen (*e.g.*, green manure, legume residues). A fertilizer rich in nitrogen therefore favours the bacterial community in a soil whereas a substrate with a relatively wide C:N ratio enables growth of the fungal population.

The activity of the soil microorganisms also shows strong seasonal variation. Activity increases markedly with increasing temperature and soil moisture. Usually fungi depend on a sufficient amount of water in the soil and are expected to be less active under dry conditions. In many cases a low pH is associated with fungal dominance whereas a higher pH might be related to bacterial dominance.

DEVELOPMENT AND PREVENTION OF PLANT DISEASE

Kinds of Organisms Cause Plant Disease

Organisms that cause plant diseases are called pathogens. They include microorganisms such as fungi, bacteria and viruses, some species of protozoas and nematodes. Pathogenic organisms are usually a normal component of the soil population and naturally exist in relatively low numbers. Some pathogenic species only cause disease in one species of plant, but others can cause disease on plants that are not closely related to each other.

Identification of Pathogen

Koch was a scientist who, in the 1880's, proposed a method for identifying the organism responsible for a disease. When his procedure is applied to identifying plant diseases the following must occur for the identification to be successful:

- An organism is isolated from a plant showing symptoms of the disease.
- The organism is grown separately from other organisms and the host (on an artificial food source).
- The organism is placed into contact with a healthy plant and the plant develops the same symptoms of the disease.
- The organism is isolated from the second diseased plant.

Soil Conditions Allow Plant Disease to Develop

Pathogens can exist in the soil for long periods of time without causing an outbreak of disease in plants. Disease outbreaks are either caused by an increase in the population of the pathogen or by an increase in the susceptibility of the plant. The population of the pathogen is dependent on whether the soil conditions are favourable for its growth and survival. The conditions that are favourable for the growth and survival of pathogens are different for each species of pathogen but are related to:

- Soil pH,
- Water content,
- Oxygen level,
- Nutrient level and
- The activities of other soil organisms.

Management practises that produce soil conditions that are unfavourable to pathogen growth will reduce the likelihood of disease outbreaks.

The susceptibility of the plant to disease is affected by factors such as its age and nutritional status. Outbreaks of disease are also more likely in agriculture and horticulture than they are in natural systems. In agriculture and horticulture similar species are planted together in what is called a 'monoculture'. Monocultures can increase the probability of a disease outbreak occurring.

Fungal and Bacterial Plant Pathogens affected Plant Roots

To enter a plant root the bacteria or fungi must first be present in the rhizosphere of the plant. It then uses molecular signals to recognise whether the plant root is susceptible to entry or not. The pathogen attaches to the root surface possibly by the use of hair-like structures and enters the root. Some pathogens enter through areas that have been damaged by animals and some fungi and bacteria produce enzymes that dissolve the chemical compounds that make up the cell wall. Once the pathogen is in the plant cells, the plant may try to prevent its spread by producing chemical or physical barriers. These procedures may confine the pathogen to a portion of the root.

Contrel of Nematodes Attack

The presence of host plants and disturbance of their environment generally stimulates the activity of nematodes. Parasitic nematodes prefer to attack the young growing roots and any areas of roots already damaged by other organisms.

Plants react in three different ways to nematode attack:

1. By production of a local swelling and 'stubby root' symptom and the immediate suppression of root growth
2. By cell death at sites where wounds and lesions have occurred
3. Through the formation of root galls resulting from cell hypertrophy

Once nematode populations have developed at a site where a particular host plant has been cropped, they may survive as resistant cysts for periods of up to 10 years. Population density generally increases steadily for five years and then stabilises. If the numbers attained at that stage are detrimental to crop yields, the usual practice is to stop growing that particular crop and either leave it to fallow or grow non-susceptible crops until the cysts are eliminated.

Biological Control

Plant disease can be controlled in a number of ways. Chemical control uses fungicides and other pesticides. Biological control of disease refers to the use of living organisms to control the numbers and activity of a pathogenic organism. These beneficial organisms may occur naturally in the soil. The objective is to manage soils so that the soil environment is suitable for growth of the beneficial organisms that naturally control plant pathogens.

SOIL BIOLOGICAL FERTILITY

Soil fertility is the combined effects of three major interacting components. These are the chemical, physical and biological characteristics of the soil. The physical and chemical characteristics of soil are far better understood than that of the biological component. It is still difficult to define the desired biological status of soil because of its dynamic nature and changes occur in much shorter time periods than physical and chemical changes in the soil.

Some aspects of soil biology may be able to provide early warnings of land degradation, such as determining the amount of carbon in microbial pools as opposed to the total amount of carbon in the soil. The biology of soil is complex and we need a better understanding of the mediatory effect that biological components have on chemical and physical fertility. We have yet to determine desirable levels of activity, numbers and diversity of soil organisms to maintain a fertile and productive soil. Since, these will change depending on the different type of soils, single optimal values will not be useful.

Ten Key Principles of Soil Biological Fertility

1. Soil organisms are most abundant in the surface layers of soil
2. Soil organic matter is necessary for nutrient cycling and soil aggregation
3. Maximum soil biological diversity depends on the diversity of organic matter and habitats
4. Nitrogen fixing bacteria form specific associations with legumes under specific conditions
5. Nitrogen is released during organic matter breakdown, either into soil or into the soil microbial biomass,
6. Arbuscular mycorrhizal fungi can increase phosphate uptake into plants in P-deficient soils
7. Soil amendments can alter the physical and chemical environment of soil organisms
8. Some crop rotations and tillage practices decrease the suitability of soil for plant pathogens
9. Production systems based on soil biological fertility can be profitable
10. Soil biological processes develop slowly, and the time required will differ for different soils, environments and land management practices.

Land Management Guidelines Relating to these Ten Key Principles of Soil Biological Fertility

1. Soil erosion should be controlled to minimise loss of soil organisms
2. Plant organic matter should be retained to maximise nutrient cycling and soil aggregation processes
3. Some disturbance of soil is necessary to maximise soil biological diversity
4. Nitrogen fixing bacteria should be selected that match the host, soil characteristics (such as pH) and environmental conditions
5. Inputs of nitrogen fertilizer should be calculated to complement nitrogen cycling from organic matter
6. Inputs of phosphorus fertilizer should be calculated to complement and enhance the activities of arbuscular mycorrhizal fungi
7. Any substance added to soil should be assessed in terms of its effects on soil biological processes and soil biological diversity

8. Crop rotations and tillage practices should be selected to avoid development of soil conditions that enhance the growth and survival of plant pathogens
9. The capacity of a management practices to produce a commercial product should be considered in parallel with its capacity to maintain and/or increase soil biological fertility
10. Sufficient time should be allowed for establishment or restoration of a level of soil biological fertility appropriate for particular soils and land management.

8

Soil Fertility and Soil Mineralogy

Soil mineralogy is closely related to soil fertility. Differences in soil mineralogy cause great differences in soil fertility. For instance, moderately weathered clays attract and retain greater amounts of nutrients than highly weathered clays and oxides. Though volcanic ash inherently has a low ability to hold nutrients, it interacts with organic matter to produce very fertile soils. Knowledge of mineralogy helps to determine the appropriate nutrient management strategy for your soil.

In our discussion on soil texture and structure, we mentioned that the very small particles form aggregates. The major groups of small particles include silicate clays, organic matter, volcanic ash minerals, and oxides. In fact, this grouping of small particles is also used to describe the distinct categories of soil mineralogy (types of clay minerals).

Soil mineralogy:

- Layered silicate clays
  - High activity
  - Low activity
- Organic matter
- Volcanic ash material
- Oxides

Soil behaviour is greatly influenced by these types of clay minerals and/ or the amount of organic matter that a particular soil type contains.

Before discussing differences in soil mineralogy, it is very helpful to understand the concepts of cation and anion exchange capacities (CEC and AEC). CEC and AEC are properties that can help differentiate soil minerals.

CATION EXCHANGE CAPACITY (CEC) AND ANION EXCHANGE CAPACITY (AEC)

CEC AND LESS WEATHERED SOILS

Less weathered soils, that contain minerals such as montmorillonite, are said to have a 'cation exchange capacity,' or CEC, under acidic, neutral, and alkaline conditions. CEC is the soil's ability to attract, retain, and supply nutrients, such as calcium, potassium, ammonium, and magnesium. These nutrients are positively charged atoms, known as cations. The surfaces of less-weathered clay minerals (such as montmorrillonite) generally have a negative charged. Much like a magnet, negatively charged soil surfaces attract positively charged cations. However, under acidic conditions, the soil will also have a tendency to attract aluminum and hydrogen cations. The presence of aluminum and hydrogen contributes to soil acidity. In contrast, under alkaline conditions, the soil attracts sodium which contributes to soil alkalinity.

AEC and Highly Weathered Soils

Although the majority of the world's soils have CEC, the highly weathered soils of the tropics are an exception. In addition to the having CEC, many tropical soils also have an 'anion exchange capacity,' or AEC, depending upon the pH of the soil. Under neutral and alkaline conditions, the soil has CEC, like the less weathered soils. However, under acidic conditions, these soils generate AEC. This means that the soil becomes positively charged and attracts, retains, and supplies negatively charged anions, such as sulfate, phosphate, nitrate, and chloride. For soils with AEC, proper management of pH is crucial in order to provide sufficient amounts of the nutrient cations (calcium, magnesium, ammonium, and potassium).

Minerals that exhibit AEC are highly weathered kaolinite, aluminum and iron oxides, organic matter, and the allophanes and imogolites of volcanic soils. Highly weathered Ultisols and Oxisols, volcanic Andisols, and organic Histosols all have AEC under acidic conditions. The pH at which these soils develop AEC differs depending upon the minerals within the soil. Since organic matter only generates AEC at a very low pH, it is still a good source of CEC.

Importance of CEC and AEC

CEC and AEC values are important measurements that provide us with important information regarding the soil's ability to retain and supply certain nutrients to the plant. In addition to nutrient retention, CEC and AEC helps us predict the leaching potential of certain nutrients in areas with high rainfall. When the soil has a very high CEC, negatively charged nutrients such as nitrate are not be retained by the soil. Instead, nitrate leaches through the soil profile in areas with high amounts of precipitation. Likewise, soils with high AEC experience leaching of positively charged nutrients, such as calcium and potassium.

CEC is often expressed in centimoles of charge per kg of soil. By sending a sample of your soil to a soil testing laboratory, you can determine your soil's CEC. The value of knowing a soil's CEC cannot be underestimated in nutrient management. Without it, your soil would not be able to provide your plants with sufficient amounts of nutrients. However, one must be cautious when interpreting the laboratory results for CEC depending on the soil type being analysed. You can obtain accurate results for most laboratory methods that measure CEC in less weathered soils with permanently negative charge. On the other hand, these laboratory methods can overestimate the CEC of highly weathered soils that have an AEC. This is because the pH of highly weathered soils affects the CEC of the soil. As the pH of highly weathered soils with AEC increases, the CEC of the soil also increases. Thus when the method used to determine the CEC uses solutions that raise the pH of the soil, the reported CEC is higher than its actual value in the field.

Layered Silicate Clays

Layered silicate clays are secondary minerals that have formed as the result of weathering of parent material. There are two major categories of layered silicate clays within the soil: high activity clays and low activity clays.

High Activity Clays

Generally, soils with large amounts of high activity clays are not highly weathered. High activity clays have a high 'cation exchange capacity' (CEC), due to their large surface area. This means that these clays have a great capacity to retain and supply large quantities of nutrients, such as calcium, magnesium, potassium, and ammonium. Not only do these clays have a large CEC, but they will generate CEC under all soil conditions regardless of soil pH. As a result, these clays tend to produce highly fertile soils. Examples of these clays are montmorillonite (and other smectites), vermiculite, illite, and mica.

Cation Exchange Capacity or Anion Exchange Capacity

- High activity clays have a cation exchange capacity (CEC). Although high activity clays will not have anion exchange capacity (AEC), the CEC increases as pH increases and decreases as pH decreases.

Montmorillonite

Some high activity clays, such as montmorrillonite, have a shrink and swell potential. This means that the clays will shrink and crack when dry, and expand and swell when wet. With little additions of nutrients, these soils may be very productive. However, the shrink and swell potential will result in poorer drainage. And so, proper management of irrigation is required.

Low activity clays

In contrast, low activity clays are more highly weathered. Thus, due their lesser surface area, low activity clays have a lower capacity to retain and supply nutrients. In addition to CEC, low activities clays can also have AEC, depending upon the pH of the soil. The AEC causes these clays to retain and supply nutrients, such as phosphate, sulfate, and nitrate, rather than the base cations, under acidic conditions. Yet, under neutral and alkaline conditions, these low activity clays generate a CEC.

Cation Exchange Capacity or Anion Exchange Capacity?

- Under acidic conditions, low activity clays have an AEC
- Under neutral and alkaline conditions, low activity clays have a CEC

Management of pH

In order to provide adequate amounts of the base cations, proper management of pH is crucial. If soil pH is low, additions of lime and/or organic matter may increase the CEC for soils high in low activity clays.

Low activity clays have a low shrink and swell potential. With additions of nutrients, these soils may be very productive soils.

Table. CEC and Surface Area of Common Soil Minerals

Mineral	**Type**	**CEC (surface charge cmolc/kg^{-1})**	**Surface area (external m^2/g^{-1})**
Smectite	High activity clay	–80 to -150	80 to 150
Vermiculite	High Activity clay	–100 to -200	70 to 120
Fine Mica	High activity clay	–10 to -40	70 to 175
Chlorite	High activity clay	–10 to -40	70 to 100
Kaolinite	Low activity clay	–-1 to -15	5 to 30
Gibbsite	Al-oxide	+10* to -5	80 to 200
Goethite	Fe-oxide	+20 to -5	100 to 300
Allophane	Amorphous	+10 to -150	100 to 1000
Humus	Organic	–100 to -500	Variable

Organic Matter

Most soil organic matter accumulates within the surface layer of the soil. This organic matter may be divided into two groups: non-humic matter and humic matter.

Non-humic matter includes all undecomposed organic material within the soil. Examples of non-humic matter are twigs, roots, and living organisms. Humic matter includes humic acids, fulvic acids, and humin. (Humin is the dark material in soil that is highly resistant to decomposition.)

Importance of soil organic matter

- Due to its tremendous surface area, soil organic matter:
  - Acts like a sponge to store water
  - Retains and provides nutrients (CEC)
  - Glues and binds soil particles into stable aggregates
- Reduces the occurrence of aluminum toxicities.

Like low activity clays, organic matter may have either CEC or AEC, depending upon soil pH. However, it will rarely have AEC. In fact, the pH must fall to approximately 2.0 before it will have AEC.

Cation Exchange Capacity or Anion Exchange Capacity?:

- Soil organic matter may have both AEC and CEC. However, the charges on organic matter are dependent upon soil pH. For soil organic matter to generate an AEC, the soil pH must be 2.0.

Management

Without additions of organic matter, tillage practices will greatly reduce organic matter content in the soil. And so, no-till and minimum tillage systems with the return of organic matter to the soil are gaining favour by farmers to improve and conserve soil quality.

FERTILIZERS AND NUTRIENT MINING IN SOILS

The only answer to the "nutrient mining" problem is to bring fresh nutrients into the area from some richer source, and fortunately natural processes have concentrated the major nutrients in locations around the world. All "synthetic" nitrogen fertilizers obtain their nitrogen from the atmosphere. This is "fixed" using the *Haber process,* in which dihydrogen and dinitrogen at high temperature and pressure are combined over a catalyst, which speeds up the reaction forming ammonia:

$$N_2(g) + 3H_2(g) \rightarrow 2NH_3(g).$$

The hydrogen required for the Haber process is normally made using methane separated from natural gas. Methane is reacted with water and air at high temperatures over catalysts to form a mixture of dihydrogen, dinitrogen (from the air) and carbon dioxide. Carbon dioxide and water are removed (the carbon dioxide is used later to make urea), leaving dihydrogen and dinitrogen in the correct proportion to combine to form ammonia (unbalanced equation)

$$CH_4(g) + H_2O(g) + [O_2(g) + N_2(g)] \rightarrow [N_2(g) + 3H_2(g)] + CO_2(g).$$

The ammonia obtained from the Haber process is converted into a form more useful as fertilizer. It can be combined with carbon dioxide to form urea, which is chemically identical to the urea excreted in the urine of animal:

$$2NH_3(g) + CO_2(g) \rightarrow O{=}C(NH_2)_2(s) + H_2O(l)$$

or combined with sulfuric, nitric or phosphoric acid to form ammonium sulfate, ammonium nitrate or diammonium hydrogen phosphate $(NH_4)_2(HPO_4)$. The nitric acid is formed by oxidising ammonia over a catalyst

$$NH_3(g) + 2O_2(g) \rightarrow HNO_3(g) + H_2O(l)$$

Phosphorus

Phosphorus is a relatively abundant element in the earth's crust, and occurs naturally in marine deposits of *phosphate rock,* as calcium phosphate,

$Ca_3(PO_4)_2$, or the minerals *apatite*, $CaF_2.3Ca_3(PO_4)_2$, or *hydroxy apatite*, $Ca(OH)_2.3Ca_3(PO_4)_2$ (hydroxy apatite is the mineral that forms bones and teeth). The phosphate rock from the Pacific region is found on coral islands where flocks of sea birds roost. It is thought that the phosphate in their excreta, derived from their fishy diet, combines over time with coral to form the phosphate rock. Some phosphate rock is sufficiently soluble so that the phosphorus from the finely ground rock is available to plants. Other phosphate rock is too insoluble for direct use, and is commonly treated with sulfuric acid to make a mixture of calcium sulfate and the more soluble calcium hydrogen phosphate, called *superphosphate*.

$$Ca(OH)_2.3Ca_3(PO_4)_2 + H_2SO_4 \rightarrow CaSO_4 + Ca(HPO_4)_2$$

Phosphate rock can also be reacted with excess sulfuric acid to make phosphoric acid, H_3PO_4, which can be converted into ammonium hydrogen phosphate, $(NH_4)_2(HPO_4)$ for use in soluble fertilizers, or into sparingly soluble magnesium ammonium phosphate $Mg(NH_4)(PO_4)$ (magamp) for use as a slow-release fertilizer.

Sulfur

Sulfur is also an essential nutrient, and some areas of New Zealand are sulfate deficient. Use of superphosphate, or ammonium, potassium or calcium sulfate (gypsum) as fertilizer supplies the needed sulfate. The sulfuric acid needed to make superphosphate is obtained by burning sulfur to form sulfur dioxide,

$$S(s) + O_2(g) \rightarrow SO_2(g)$$

which is oxidised over a catalyst to form sulfur trioxide,

$$2SO_2(g) + O_2(g) \rightarrow SO_3(g)$$

which reacts with water to form sulfuric acid,

$$SO_3(g) + H_2O(l) \rightarrow H_2SO_4(aq).$$

The sulfur is obtained from deposits of elementary sulfur found in volcanic areas of the world, or is extracted from natural gas, which in some places contains hydrogen sulphide. Sulfur dioxide is released into the atmosphere from volcanoes. This is ultimately converted into sulfuric acid in the atmosphere; this is washed out by rain, so providing a "natural" source of sulphur.

The 1995 eruption of Ruapehu "topdressed" much of the central North Island with economically useful amounts of sulphate. Sulfur dioxide is also released into the atmosphere when coal or oil containing sulfur is burned in power stations, and the resulting sulfuric acid in the atmosphere causes "acid rain" in some industrialised areas.

Potassium

Potassium is the other main nutrient needed, and this is obtained from sea water, directly or indirectly. When sea water evaporates sodium chloride (common salt) crystallises first. If the liquid remaining (bittens) is evaporated potassium chloride crystallises. In some places this occurred over geological times when oceans were isolated by geological processes and then evaporated, leaving large deposits of salt (which crystallised first) and of potassium chloride (which crystallised later). The potassium chloride can be used directly as a fertilizer, and is satisfactory for use at low rates, but this has the disadvantage that chloride is also added, and many plants are not very tolerant of high levels of chloride.

It is therefore usual to convert the potassium chloride to potassium sulfate (which also occurs in natural deposits) for use as a fertilizer. The so-called "un-natural" inorganic fertilizers are thus derived from the very natural sources of the air (plus natural gas as a hydrogen source) for nitrogen, coral rock plus sea-bird droppings (plus sulfur from volcanoes or "sour" natural gas to make superphosphate) for phosphorus, and the ocean for potassium.

MAINTENANCE OF SOIL FERTILITY

No two soils are alike either in respect of their nature or in respect of quantities of plant nutrients they contain. Under a given situation, the system of farming, soil management and manuring practices, etc., influence the productiovity of soils and crop yields obtained from them.

It is estimated that the different agricultural crops in India remove about 4.27 million tonnes of nitrogen, 2.13 million tonnes of phosphoric acid, 7.42 million tonnes of potash and 4.88 million tonnes of lime per year. The production of larger yields through improved varities of crops and intensive cultivation will increase the depletion of nutrients still further. But erosion and leaching cause additional losses. The present production of synthetic nitrogenous fertilizers in the country reached 1.5 million tonnes of nitrogen, and the bulky organic manures might supply another 1.5 million tonnes of total nitrogen.

The amounts of phosphorous, potash, etc., added to the soil are very small. It is thus obvious that the current huge drain on nutrient supplies will continue to impoverish the soils unless these supplies are replenished by natural or by artificial means, the principal methods of supplementing natural recuperation and for improving the productive capacity of the soils are:

- To add organic matter to the soil, so that through decay, it may furnish a more or less continuous supply of nuttrients for crops,
- To restore or increase the amount of defficient nutrients by the application of fertilizers.

The urgent need of the constantly expanding agriculture production to meet the requirements of the continually increasing human and cattle populations in India makes the supply of additional plant nutrients through fertilizers and organic manures, a problem of supreme importance.

TYPES OF MANURES AND FERTILIZERS

Indian soils are usually very poor in organin matter as well as in nitrogen. Pohsphate deficiency is less wide-spread and potash deficiency generally occurs in com-pact areas. In acid soils the addition of lome steps up production.Materials which are commonly used to maintain and improve soil fertility

- *Manures*: These are relatively bulky materials, such as animal or green manures, which are added mainly to improve the physical condition of the soil, to replenish and keep up its humus status, to maintain the optimum conditions for the activities of soil micro-organisms and make good a small part of the plant nutrients removed by crops or otherwise lost through leaching and soil erosion.
- They, thus, supply practically all the elements of fertility which crops require, though not in adequate proportions. The plant-food elements contained in a manure are released in an available form after it is applied to the soil and is decomposed by soil micro-organisms. Similarly, the green manures add not only substantial amounts of organic matter but also nitrogen.
- *Fertilizers*: Fertilizers are inorganic materials of a concentrated nature; they are applied mainly to increase the supply of one or more of the essential nutrients, e.g. nitrogen, phosphorous and potash. Fertilizers contain these elements in the form of soluble of readily available chemical compounds. This distinction is, however, not very rigid. In common parlance the fertilizers are sometimes called 'chemical','artificial' or 'inorganic' manures.
- *Concentrated Organic Manures*: Some of the concentrated materials, such as oil-cakes, bone-meal, urine and blood are of organic origin. The use of manures and fertilizers is complementary and not as a substitute for each other.
- *Bulky Organic Manures*: The properties and role of organic matter and humus in the soil have been explained already.

Farmyard Manure

Good-quality farmyard manure is perhaps the most valuable organic matter applied to a soil. It is the most commonly used organic manure in India. It consists of a mixture of cattle dung, the bedding, used in the stable and of any ramnants of straw and plant stalks fed to cattle. Though its crop-increasing value has been recognised from time immemorial, more than 50 per cent of the cattle dung produced in the country today is burnt as fuel and is thus lost to agriculture. Not only this tremendous waste, but also the tradition method of preparing and storing the farmyard manure is generally faulty. The cattle-dung, together with stable-waste and house sweeping, is first collected in the open backyard, and when a cartload has been collected, it is removed to another heap or to an uncovered pit in a common plot outside the village. The loose heaps lie exposed to the sun, with the result that the raw organic matter dries up quickly and does not rot properly.

Very often, a part of the dry dung is blown off by wind or washed away by rain. Cattle urine is either not conserved or is stored in a defective manner. American studies on the distribution of soil derived elements between urine and faeces of dry cows have shown that 95 per cent of Potassium, 63 per cent of nitrogen and 50 per cent of sulphur are contained in the urine.

The wastage of nitrogen-rich urine, the loss of nitrogen(in the form of ammonia) due to the fermentation of exposed cattle dung, and the washing away of soluble mineral elements be leaching reduce its manurial value in India to a great extent. Its average content of plant nutrients under Indian and European conditions is shown below for comparison:

Percentage Content

	N	P_2O_5	K_2O
India	0.3	0.15	0.3
European countries	1.0	0.30	1.0

About half of this nitrogen, one-sixth of phosphorous and more than half of potash are readily soluble and subject to dissipation. However, the loss of nitrogen and minerals elements caused by careless handeling can be reduced greatly by using absorbent bedding for cattle, storing dung in stone or brick-line pits, mixing large quantioties of straw and other vegitable matter with cattle dung, and keeping the heap compact and moist.

Thus, if urine is properly conserved, the loss of soluble mineral elements through seepage is prevented, bacterial decomposition of raw organic matter is encouraged, plant nutrients are made soluble, and nitrogen losses are minimised.

Table. The Relative Absorbent Capacity of Different Materials used as Bedding for Cattle

Material	Quantity of water (in kg) retained by one kg of the following materials after 24 hours of soaking
Wheat straw	2.20
Peat straw	2.80
Dry leaves	2.00
Peat	6.00
Sawdust	4.35
Soil	0.50
Sand	0.25

If urine is not conserved in the bedding used for cattle it must be collected in covered *pucca* cistern and then, added to the dung in the manure pit. Nitrogen in the urine is mainly in the form of urea, which readily changes into the highly volatile ammonium carbonate through bacterial action, and quickly loses ammonia thereafter by evapouration. This loss can be reduced to a great deal if the manure and the urine-soaked absorptive itter for bedding are kept compacted in a pit.

The pit may be 1 m in depth, 1.3 to 1.5 m in width and 4.5 to 6 m in length, depending upon the no. of cattle on a farm. The filling of the pit should be 'sectional' and when each section of three or 1.3 m in length is filled to about 45 cm above the ground level, it should be clustered with 2.5 cm layer of a mixture of mud and dung in equal proportions. Before plastering, 4 to 5 buckets of water should be added to the manure in the pit. Plastering conserves moisture and nitrogen and also prevents housefly nuisance. The manure becomes ready for use in about 4 to 5 months after plastering.

The quality of manure is also improved by the concentrated feeds given to cattle. Cotton-seed, cotton-seed cake, linseed-meal, wheat bran, grain husk, groundnut cake, gram, horse-gram, etc. are rich in nitrogen, phosphorous, potassium, magnesium and sulphur.

It has been found that in the case of adult working-cattle about 80 per cent of nitrogen and the other mineral elements contained in the feed is recovered in urine, faeces and other animal by-products. Accordingly, manure from cattle fed on cereal straws and grass hay is much less valuable than that from animals fed on legume hays, grains and concentrates.

In foreign countries, considerable attention has been given to the use of preservatives on manure. Calcium sulphate or gypsum and superphosphate have proved most promising in preventing the escape of ammomnia. Gypsum has been found specially effective as an ammonia-absorbing agent. Superphosphate, besides absorbing ammonia, supplies additional phosphorous and, thus, improves the crop producing capacity of the manure.

Partially rotten farmyard manure should generally be applied to the soil about three to four weeks before the sowing of a crop. In case there is sufficient moisture in the soil, there sill be enough time for its decomposition and for improving the soil structure. Its application too long before sowing a crop will either cause a drying up of the rotted manure or too quick decomppostion, depending on the incidence of rains. But in each case, there will be a serious loss of ammonia and nitrogen. If the manure is already well rotted, it is advisable to apply it just before sowing to a crop.

This procedure is perticularly essential in the case of light soils. In any case, after the manure is carted to the field, it should be evenly spread and worked into the soil soon to avoid the loss of nitrogen. The existing practice of leaving the manure in small heaps scattered in the field for several days, before spreading it in the field and incorporating it into the soil, results in the serious deterioration of its quality, particularly if strong winds blow. In vagitable and fruit cultivation, the application of well-rotten manure, in conjunction with fertilizers to young plants individually, has been founnd to give the best results. In Egypt, even the cotton crop, which is invariably grown on ridges, is given a top-dressing a handful of the rotten manure being applied to the soil at the base of each plant before an irrigation, and is worked into the soil with a hand hoe. The practice of penning cattle, sheep and goats in the fields in summer is common in some parts of the country. Folding 7,000 sheep for one night is said to add the equivalent of 149.3 quintals(14.93 tonnes) of cattle dung. The fresh dung left in the field in such cases rapidly dries up. This drying checks ammonification and loss of nitrogen. With the first fall of rain, the dung is worked into the soil. It, therefore, does not lose much of its fertilizing value. Further more, its beneficia effects on the physical condition of the soil are undeniable. However, sheep-folding is said to make the land more weedy.

There must be adequate moisture in the soil for the proper decomposition of organic matter. Farmyard manure can, therefore, be applied to all crops grown in the rainy season or grown under irrigation. The quantity of manure to be applied to unirrigated crops varies from 1.5 to 2 cartloads per hectare in areas of heavier rainfall. If sufficient farmy manure is not available, it may be applied at the usual rate to a part of the land,say, to one-third or one-forth of the area, in rotation every year, so that all parts of the field receive the manure regularly once in three or four years. For irrigated field crops, the rate varies from 10 to 20 cartloads. Sugarcane, maize and garden crops, such as potatoes, turmeric, ginger, vegetables and fruits receive still higher doses, amounting sometimes to 15 to 25 cartloads. A cartload of manure, measuring 9 cubic metres, weighs about half a tonne.

It must be stressed that the value of farmyard manure in soil improvement is due to its content of principal nutritive elements and its ability to:

- Improve the soil tilth and aeration,
- Increase the water-holding capacity of the soil,

  Stimulate the activity of micro-organisms that make the plant-food elements in the soil readily available to crops.

The supply of organic matter, which is later converted into humus is a property of farmyard manure. One tonne of carttle dung can supply only 2.95 kg of nitrogen, 1,59 kg of phsophoric acid and 2,95 kg of potash.

The use of farmyard manure alone causes an imbalance in nutirtion owing to its relatively low content of phosphoric acid. Therefore, to keep the soils well supplied with all the essential elements of plant food in a readily available form, and also to keep them in good 'heart', it is advisable to use the bulky organic manures in conjuction with superphosphate and such other artificial fertilizers as contain the particular plant food or foods in which a soil may be deficient or which the crop to be grown may specially require.

9

Soil and Water Conservation Measures

The key to soil and water conservation is the utilisation and treatment of land according to its capability.

Land-Capability Classification

Any soil and water-conservation project includes two distinct sets of operations, viz.

1. The mapping of land for classification according to its capability,
2. Planning and executing measures to check erosion, improve land productivity and reclaim wasteland.

The farm plans for effective soil and water conservation are based largely on the capability of the land. The land-capability classification map is normally prepared by interpreting a standard soil-survey map. Land-capability classification is a systematic arrangement of different kinds of lands according to those properties that determine the ability of the land to produce crops on a virtually permanent basis.

The Factors Determining Land-capability

These are the major soil characteristics of the land, *e.g.,,* the texture of the top soil, its effective depth, permeability of the top soil and subsoil, and associated land features, *e.g.,,* the slope of the the land, the extent of erosion, the degree of wetness and susceptibility to overflowing and flooding.

The grouping of soils into capability classes is done primarily on the basis of their capability to produce common cultivated crops and pasture plants without deterioration over a long period.

Land-capability Classes

The land-capability classes are based on the intensity of hazards and the limitations of use. The land-capability classes range from the best and most easily farmed land to that which has no value for cultivation, grasing or forestry, but which may be suited to wild-life, recreation or for watershed protection. They all fall into 2 broad groups: one suitable for cultivation and other land uses, and the other not suitable for cultivation, but suitable for other land uses.

Land Suitable for Cultivation and Other Uses

Class I (Green Colour)

Soils in class I have very few or no limitations that restrict their use.

This type of land is nearly level and the erosion hazard is low. The soils are deep, well-drained, easily worked, hold water well and are either fairly well supplied with plant nutrients or highly responsive to the application of fertilizers. The soils are not subject to damage because of overflow. The local climate must be favourable for growing many of the common field crops. In irrigated areas, the soils may be in class I, if the limitation of the arid climate has been removed by relatively permanent irrigation works.

These soils need ordinary management practices to maintain productivity. Such practices may include the use of one or more of the following: fertilizers, lime, cover and green-manure crops, conservation of crop residues and crop rotations.

Soils in this class are suited to a wide range of plants, may be used for cultivated crops, pastures, forests, and wildlife, food and cover.

Class II (Yellow Colour)

Soils in class II have some limitations which reduce the choice of plants or require simple conservation practices.

The limitations of soils in class II may result from the effects of one or more of the following factors:

- A gentle slope,
- A slight susceptibility to erosion,
- Less than ideal soil depth,
- Occasional damaging overflow,
- Wetness which can be corrected by drainage, but existing permanently as a moderate limitation,
- Slight to moderate salinity or sodium, easily corrected but likely to recur,
- A slight climatic limitation on soil use and management.

These soils require careful management. The limitations are only a few and the practices are easy to apply. They may need one or more of the following practices: terracing, strip cropping, contour cultivation, water disposal area, covered with vegetation crop rotation, cover and green-manure crops, stubble mulching, the use of fertilizers, manure and lime. These soils may be used for growing cultivated crops, raising pastures, forests, and for wild-life, food and cover.

Class III (Red Colour)

Soils in class III have moderate limitations which reduce the choice of plants or require special conservation practices. Soils in class III have more restrictions than those in class II and, when used for cultivated crops, the conservation practices are usually more difficult to apply and to maintain.

Limitations of soils in class III may result from the effects of one or more of the following factors:

- A moderately sloping land,
- Moderately susceptibility to water or wind erosion,
- Frequent overflow accompanied with some crop damage,
- Very slow permeability of the sub-soil,
- Wetness or continuing water-logging after drainage,
- Shallow soil depth up to the bed-rock, hard-pan or clay-pan which limits the rooting-zone and water storage,
- Low moisture-holding capacity,
- Moderate salinity or sodium,
- Moderate climatic limitations.

Class IV (Blue Colour)

Soils in class IV have severe limitations that restrict the choice of plants and require careful management. The restrictions in the use of these soils are greater than those in class III and the choice of plants is more limited. When these soils are cultivated, very careful management is required and the conservation practices are more difficult to apply and to maintain.

The use of these soils for cultivated crops is limited as a result of the effect of one or more of the permanent features, such as:

- Steep slopes,
- Severe susceptibility to water or wind erosion,
- Severe effect of past erosion,
- Shallow soil,
- Low moisture-holding capacity,
- Frequent overflow accompanied with severe crop damage,
- Excessive wetness or continuing hazard of water-logging after drainage,
- Severe salinity or sodium,
- Moderately adverse climate.

These soils can be used for crops, pastures, forests, and wild-life food and cover.

Land Not Suitable for Cultivation but Suitable for Other Land Uses

Class V (Dark Green or Uncoloured)

Soils in class V have little or no erosion hazard, but have other limitations, the removal of which is not practicable. They are used largely for pastures, forests, and wild-life food and cover. Such land is nearly level and is not subject to more than slight wind or water erosion. Cultivation is not feasible because of one or more limitations, such as overflow, stoniness, wetness or severe climate.

Examples of class V land are:

- Soils of lowlands subject to frequent overflows which prevent the normal production of cultivated crops,
- Nearly level soils with growing season that prevents the normal production of cultivated crops,
- The level or nearly level stony or rocky soils,
- Ponded areas where drainage for cultivated crops is not feasible but where soils are suitable for grasses or trees.

Soils in class V are not suitable for raising cultivated crops, but are suitable for perennial vegetation (grasing and forestry, with few or no limitation). Pastures can be improved, and benefits from proper management can be expected. Physical conditions of soils are such that it is practicable to apply pasture improvements, if needed, such as seeding, liming, fertilizing, and water control with contour furrows, drainage ditches, diversions of water spreaders.

Class VI (Orange Colour)

Soils in class VI have severe limitations that make them unsuitable for cultivation and limit their use largely for pastures, or forests, or wild-life food and cover. Soils in class VI have continuing limitations which cannot be corrected, such as:

- Steep slope,
- Very severe erosion hazard,
- Very severe effect of past erosion
- Stoniness,
- Shallow rooting-zone,
- Excessive wetness or overflow,
- Low moisture capacity,
- Salinity or sodium,
- Severe climate.

Soils in this class are subject to moderate limitations under grasing or forestry use.

Class VII (Brown Colour)

Soils in class VII have very severe limitations that make them unsuitable for cultivation and restrict their use largely to grasing, or forestation, or wild-life food and cover. The soils in this class are subject to severe limitations or hazards under either grasing or forestry use. The physical condition of soils is such that it is not practicable to adopt pasture improvements and water-control practices. Soil restrictions are sever e than those in the case of class VI soils.

Class VIII (Purple Colour)

Soils and land forms in class VIII have limitations that preclude their use for commercial plant production and restrict their use to recreation, wild-life food and cover or to water-supply, water shed protection or for aesthetic purposes. Significant return on site benefits from soils and land forms in class VIII cannot be expected from management of crops, grasses or trees, although indirect benefits from wild-life, watershed protection or recreation may be possible.

Limitations which cannot be corrected may result from the effects of one or more of the following factors:

- Erosion or erosion hazard,
- Severe climate,
- Wet soil,
- Stones,
- Low moisture capacity,
- Salinity or sodium.

Bad lands, rock outcrops, sandy beaches, marshes, deserts, river wash, mine tailings and other nearly barren lands are included in class VIII in order to protect other more valuable soils to control water or for wild-life or for aesthetic reasons. The land-capability class is indicated on the maps by roman numerals I to VIII or by standard colours or by both.

Land-capability Subclass

A land-capability class is determined by the degree of limitations in land use together with the hazards involved. For example, in class III land, we have land suitable for cultivation but subject to moderate hazard of water erosion because of a steep slope or a moderate hazard of wind erosion on smooth land or moderate hazard of water-logging or overflow and a shallow depth to the bed-rock. Each of these kinds of limitations are recognised at the subclass. Four kinds of limitations are recognised at the subclass level.

Subclass (e) Erosion

It is made up of soils where the susceptibility to erosion is the dominant hazard in their use. Susceptibility to erosion and damage due to past erosion are major factors that govern the placing of soils in this class.

Subclass (w) Excess Water

It is made up of soils where excess water is the dominant hazard in their use. Poor soil drainage, wetness, high water-table and overflow are the criteria for determining which soils belong in this subclass.

Subclass (s) Soil Limitations within the Rooting Zone

It includes soils which have such limitations as the shallowness of the rooting-zone, stones, low moisture-holding capacity, low fertility difficult to correct and salinity or alkalinity.

SOIL WATER RELATIONS UNDER FIELD CONDITIONS

In a region of moist climate, if a hole be dug or bored into the ground at almost any place where the soil is deep enough a level at which the soil is completely saturated with water will be encountered at some depth or other. Water will stand in this hole up to this level of complete saturation which is called the water table. In river valleys or in close proximity to large bodies of water the water table will usually be reached at a depth of only a few feet under the soil surface. Even in regions of arid climate there are some local situations in which a water table is present. The depth of the water table in any locality usually fluctuates, sometimes very markedly, with seasonal and periodic changes in the relative rates of rainfall, evapouration, transpiration and other factors. Relatively impermeable soil layers sometimes impede downward percolation of water sufficiently to cause the development of temporary water tables which may be far above the level of the true water table. Such temporary water tables have essentially the same effects on soil water relations as true water tables, except that their influence is often only a transient one.

For many years an important role was ascribed to the capillary rise of the water from the water table into the soil above in maintaining the moisture conditions within that soil. Recent investigations have shown, however, that the importance of this source of water in soils has been overemphasised. Experiments upon the rise of water through columns of soil indicate that water seldom rises through the soil by capillarity from the water table at an appreciable rate for heights of more than a few feet. In a typical loam soil the absolute maximum capillary rise of water is about 8 feet. Such capillary rise as does occur takes place most slowly, but to the greatest height in clay soils, and most rapidly but to lesser heights in sandy soils, loam soils occupying an intermediate position.In many soils the water table lies so far below the soil surface that it has little or no effect on the soil moisture conditions in the soil layers which are penetrated by the roots of most plants. Generally speaking even in loam or clay loam soils, a capillary rise of water from a zone of complete saturation is probably ineffective in providing the roots of most species of plants with any considerable part of the water which they absorb unless the water table is within about 15 feet of the surface. Some of the more deeply rooted species may obtain some water from a water table located at depths as great as about 30 feet, but the presence of a water table at greater depths than this is an egligible factor in supplying the roots of any species of plants with water.

A capillary rise of water from a water table is usually an important source of water for plants in such locations as river bottom lands or in fields or forests in close proximity to ponds or lakes. In many and perhaps most agricultural soils in moist climate regions plants obtain some water from the water table at least during the earlier part of the growing season.

However the widespread practice of drainage of agricultural regions has often resulted in such a marked lowering of the water table as to greatly reduce the possibility that capillary rise of water from below will supply crop plants with any significant part of the water which they absorb. The other important source of soil water is that part of the precipitation usually rain, although often melting snow or ice, which percolates downward into the soil. In dry regions this is the only source of water available to plants.

Even in more humid regions this is the principal or only source of available soil water in most soils during the dry season of the year. Not all of the rain which falls on the surface of a soil penetrates into it. Some maybe lost by surface run-off and a portion usually evapourates before it can sink into the soil. The proportion of the precipitation which is lost in the run-off is in general much less with open, porous soils such as those found in most forests than with those of a lower porosity.

A smaller proportion of the water is lost in the run-off when it falls as a gentle rain than when it comes as sudden downpour. The proportion of the precipitation which is lost by evapouration from the surface layers of the soil depends in part on the colour, texture porosity, and other properties of the soil, and in part on the temperature and vapour pressure of the atmosphere, as well as upon the impinging solar radiation.

Let us now consider how percolating rainwater becomes distributed in as oil. We will assume that the soil under consideration is approximately air dry, that it is essentially homogeneous to a depth of several feet, and that Let us now assume that another two inches of rain falls on this same soil before any appreciable amount is lost from its upper layers by evapouration or transpiration. After a new equilibrium has been attained the upper foot of soil will not have increased in water content, but a layer of the soil approximately two feet in depth will now be moistened up to its field capacity. In other words addition of a second increment of water equal in volume to the first does not increase the water content of the already moist layer of the soil above its field capacity, but results, due to further capillary movement, in raising to its field capacity a second layer of soil, lying directly under the zone which had been moistened by the addition of the first increment of rain, and approximately equal to it in thickness.

Although the exact mechanics of the distribution of water under these conditions is probably more complicated, the effect produced is as if the second increment of water simply flows through the moist blanket of soil, after which it is distributed to the dry soil layer underneath by capillarity. Successive rainfalls would thus continue to deepen the layer of soil which has been moistened up to its field capacity. If the water table is not too deep, and if sufficient rainfall penetrates the soil, not too much of which is utilized by plants growing thereon, eventually the entire soil from its surface down to the water table will be moistened up to its field capacity. A zone several feet in height just above the water table may be enriched somewhat above field capacity by upward capillary movement from the water table.

If additional water is applied to the soil, after the entire soil mass down to the water table has attained its field capacity, this water percolates down through the soil under the influence of gravity and becomes apart of the ground water. Such conditions obtain in many of the soils of more humid regions at least during the wetter seasons of the year, provided that the water table is not located at too great a depth. This downward percolation of water to the water table is an important factor in influencing its depth below the soil surface, which usually fluctuates considerably from season to season.

In the preceding discussion we have assumed, in the interests of simplicity a homogeneous soil. However, most soils consist of a vertical succession of several distinct horizons, each with more or less distinctive physical and chemical properties. Even after long continued disturbance of a soil by plowing or other cultural activities at least some semblance of the original soil stratification usually persists. In the horizons below those reached by the plow the original structural organisation of the soil is usually maintained practically intact. Although the individual horizons of a soil are often fairly homogeneous within themselves, each such horizon in a given soil may have a distinct field capacity of its own.

The water contents of the different horizons, even after an equilibrium in the capillary distribution of water has been attained, may therefore be very different. A somewhat erratic distribution of rainfall to an underlying soil often results from cracks or channels which are opened into the soil by one agency or another. Many soils crack upon drying, sometimes to depths of several feet. Such cracks often facilitate a mass flow of rainwater to a considerable depth in the soil. Similarly in forest soils the decomposition of roots may leave channels through which water flows down into the deeper layers of the soil. The burrows of animals may also facilitate the entry of water into a soil.The presence of rock strata, or of relatively impervious layers of soil close to the surface, will also modify very markedly the simple picture which has been presented of soil water conditions.

LABORATORY MEASUREMENTS OF SOIL WATER RELATIONS

The problem of the water relations of soils has been approached both from the standpoint of laboratory and field studies. Some of the broader generalisations resulting from field studies have been just described. Of the numerous laboratory measurements of the water relations of soils which have been devised, only a few of the better known will be discussed. Water Content. The water content of a soil is commonly expressed as the percentage of water present in terms of the dry weight of the soil. It is measured by drying a sample of the soil in an oven to constant weight at (usually) 105^{0}C. and assuming that the loss in weight represents water.

There is however nothing critical about the temperature 105^0C. The relation between the water content as determined by this method and the temperature at which the soil is dried is a linear one over a wide range of temperatures extending on both sides of this point. This shows that there is no especial significance to be attached to the amount of water which is lost from a soil when dried at this temperature. There is also a small error in such determinations due to some decomposition of organic matter at this temperature. Determinations of the total water content of the soil contribute very little to an understanding of the absorption of water from the soil by plants. As will be shown later the amount of water available to plants may be very different in two soils of identical water contents.

SOILS AND SOIL WATER RELATIONS

Most vascular plants are rooted in the soil from which they obtain both water and mineral salts. The entrance of water from the environment into any of the organs of a plant is called the absorption, intake, or uptake of water. In most vascular species the quantity of water absorbed through organs other than the roots is of negligible importance. Any thorough consideration of the entrance of water into roots requires a consideration of the properties of soils, particularly as they affect the movement of water from the soil into the root.

CONSTITUTION OF THE SOIL

The soil matrix which is the habitat of most roots is an extremely complex system. In general five different components of this system are distinguished:1. The Mineral Matter of the Soil.The parent substance of practically all soils is rock, of which there are numerous varieties. By various weathering processes rock strata are reduced to fragments of diverse sizes and these compose the bulk of most soils. The rock particles in soils may vary in size from stones and gravel down to sub-microscopic particles of colloidal clay.

The mineral portion of the soil is customarily classified into several fractions depending upon the size of the particles. Such classifications usually disregard the very large particles of the soil (rock fragments, pebbles, etc.). The proportions of the different fractions present are very different in different kinds of soils.The clay fraction of the mineral portion of soils requires special consideration. Unlike the coarser fractions which are composed of small fragments of unmodified rock minerals such as quartz, feldspar, and mica, the clay portion of soils is made up almost entirely of the products of chemical weathering of the rock minerals and hence differs not only in its physical state, but in its chemical composition from the particles in the coarser fractions.

Most of the particles in the clay fraction are of colloidal dimensions and exhibit the characteristic properties of colloidal systems. The particles of the clay complex, like those of many other colloidal systems, retain water by imbibition, *i.e.* within the structure of the particle. On the contrary the sand

and silt particles of the soil retain water only on their surfaces. The presence of a considerable proportion of clay in a soil therefore endows it with a high water retaining capacity. Changes in the water content of colloidal clay result in marked changes in its volume. One result of such changes in volume is the commonly observed cracking of a soil rich in clay upon drying. The plasticity and cohesiveness of soils are also due very largely to the colloidal clay present.

Like other colloidal systems colloidal clay is markedly sensitive to the influence of electrolytes. The micelles of colloidal clay usually bear a negative charge when in contact with water. In the presence of calcium ions the individual clay particles are more or less completely flocculated into compound particles. Much of the colloidal clay in most soils exists as enveloping films around the larger soil particles and is also closely associated with organic material in the soil. Flocculation of the clay particles therefore usually results in the formation of compound granules including sand and silt particles and organic matter in addition to the clay.

These are called soil crumbs. For agricultural purposes a well developed granular structure of the soil is highly desirable since such a structure favours both a high moisture-retaining capacity and a good aeration of the soil. Calcium soils, that is, those with a high calcium content, are in general the most suitable and most valuable for agricultural purposes partly because calcium favours the development of a crumb structure. In the presence of an excess of univalent cations (Na^+, etc.), on the other hand, a greater proportion of the clay fraction of the soil disperses into its ultimate particles, and the soil has a single grain structure.

In this condition the soil is in the least desirable structural condition for agricultural purposes. Many soils contain hydrogen ions in excess. Such soils often develop a good crumb structure, but the crumbs are less stable than those developed in a calcium soil. The addition of lime improves the physical structure of such soils. The crumb structure of a soil, especially in its surface layers, can also be destroyed by purely mechanical effects, such as trampling,or beating by heavy rains. The Organic Matter of the Soil.

Most soils contain organic matter which has been derived principally from the partial decomposition of plant residues. Small quantities may also originate from animal residues and excretions. The proportion of organic matter present may vary from almost none as in some sand deposits to 95 per cent or better in some peat soils. In ordinary agricultural soils the amount present seldom exceeds 15 per cent. In forests organic matter comes from falling leaves, dead branches and trunks of woody plants, roots which die and decay underground, and from dead herbaceous vegetation. In grasslands the underground roots and rhizomes as well as the aerial parts of the plants all contribute their quota to the organic matter of the soil. In well managed agricultural soils attempts are made to maintain the organic matter content by supplying them with organic fertilizers.

The organic matter is the seat of most of the microbiol-ogical processes occurring in the soil. One of the most important of these is the oxidation of the organic matter, a process due largely to the metabolic activities of bacteria and fungi, although a limited amount of purely chemical decomposition probably also occurs. Under conditions which are exceptionally favourable for the activities of microorganisms the organic matter of the soil is oxidised completely and disappears.

For this reason the organic matter of soils in tropical regions, particularly when under cultivation, is very low. Even in more temperate regions cultivation of the soils generally results in a rapid reduction inorganic matter content, due principally to the better aeration induced by till age. As a result of the decay process there is present in most soils organic matter in various stages of decomposition. A large proportion of the organicmaterial which is added to some soils survives in the form of a dark-colouredamorphous substance called humus. Humus is composed principally of the degradation products of the cellulose and lignin derived from plant remains.

The accumulation of humus in soils is furthered by conditions unfavourable to the oxidative decomposition of organic matter. The decomposition of organic matter in bogs, ponds, swamps, and water-logged soils under conditions which are largely if not entirely anaerobic results in the production of relatively large quantities of humus, frequently of the type which is called peat. Where the organic matter supplied is distinctly acid, as under coniferous forests, or heath plants, humus usually accumulates as a definite layer at the surface. Decomposition of the organic remains under these conditions is largely effected by fungi. In prairie and steppe regions, where a grassland vegetation is predominant, humus usually accumulates in considerable quantities as a result of the decay of both underground and aerial organs of plants. The amount of humus which accumulates in any soil depends upon the relative rates of the addition of organic residues and of its disappearance as a result of oxidation. Soils with a low humus content may result from a sparse contribution of organic matter, as in desert or semi-desert regions, or from a rapid oxidation of organic material which prevents accumulation of humus even when the supply of organic residues to the soil is large.

This latter condition obtains in many tropical and sub-tropical soils. Humus is essentially colloidal in its properties and possesses an even greater imbibitional capacity than the colloidal clay particles of the soil. The plasticity and cohesiveness of the colloidal organic matter, while considerably less than that of the colloidal clay, is much greater than that of the non-colloidal fractions of the soil. Humus is rather inert chemically, its influence on the soil being largely a physical one. Its presence in soils favours a looser structure and hence better aeration, a higher water holding capacity and a reduction in cohesiveness as compared with heavy clay soils.

Since the clay fraction and the humus are the two essentially colloidal fractions of the soil, and are associated in an intimate physical relationship many soil scientists refer to these two soil fractions, considered jointly, as the colloidal complex of the soil. A large part of this colloidal complex occurs in many soils, as films enveloping the larger soil particles. The relation of these films to the maintenance of the crumb structure of the soil has already been mentioned. Many other important soil properties are either due to or greatly influenced by the colloidal complex.

THE EFFECT OF WATER ON SOIL

THE TROPICAL RAIN FOREST

The lushness of the jungle biome is somewhat illusory. While productivity is high, the soils themselves tend to be of very poor quality. Because of the high rainfall, nutrients are quickly washed out of the topsoil unless they are incorporated in the forest plants. As plant and animal debris falls to the ground, it is quickly decomposed because of the warmth and moisture there.

Thus minerals are found mainly in the forest plants, not in the soil. When the plants are removed and cultivation attempted, the soils quickly lose fertility. The situation is made worse by the lack of humus (the topsoil may be no thicker than 2 in.) and the high iron and aluminum content of most of these soils. Once exposed to the sun, these lateritic soils soon bake into a bricklike material that cannot be cultivated.

The most ancient (some might say primitive) way of working these soils is still the best:

- Clearing a small area of jungle,
- Growing crops for only a year or two, and then
- Abandoning the area to jungle once again.

THE TEMPERATE DECIDUOUS FOREST

These regions receive 75–100 cm or more of precipitation each year. Enough water falls on the soil so that much of it passes down to the water table. As it does so, it carries minerals with it. Such soils tend to be acidic and of low and (if unattended) diminishing fertility. Only by regular fertilization and liming (to restore calcium and raise pH) can productive agriculture be carried out in them. In the U.S., the soils east of the Appalachian Mountains tend to be of this sort.

GRASSLANDS

In the plains of North America, the annual rainfall is sufficiently low (~50 cm) that little or no rainfall percolates down to the water table. Calcium and other minerals are not carried below the reach of plant roots and so remain available for use. This keeps the pHand general fertility high.

Except to the extent that minerals are lost when crops are removed, the minerals simply recycle from subsoil to topsoil and back to the subsoil. The self-restoring fertility of the soils of the plains states accounts for this region being the "breadbasket" of the nation.

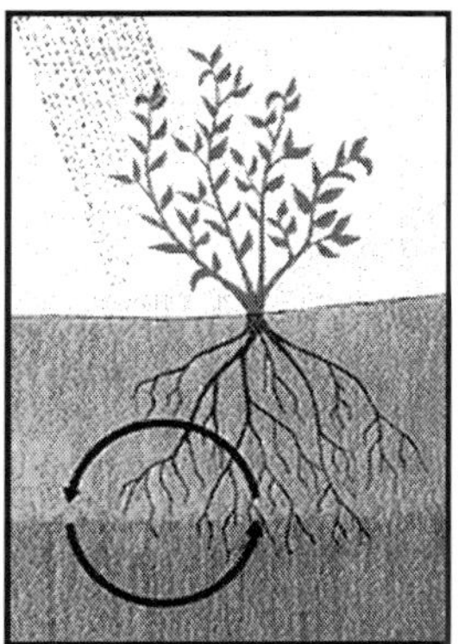

The figure shows the cycling of minerals in grasslands soil. The grasses in undisturbed prairie are perennial; their extensive root systems help prevent soil erosion, and the return of the season's above-ground growth to the topsoil returns minerals and provides humus to it. These advantages are diminished when annual grasses such as wheat and corn are grown instead and removed in the harvest.

DESERTS

The rainfall here is so low (25 cm/year or less) that any water that does not immediately run off remains near the surface and is largely lost by evapouration. The saltsit carries are left near the top of the soil. Their accumulation may make the soil so alkaline and so salty that most crops cannot be grown.

In the U. S., the situation is especially severed in the Great Basin states because water flowing down from the mountains — bearing its load of dissolved salts — cannot flow on to the ocean but simply flows out onto the valley floors and evapourates. Large areas of formerly unproductive desert in the United States, Israel, and Egypt have been converted into fertile fields through irrigation.

But irrigation is no panacea. Even the best irrigation water contains dissolved salts. If just enough water is applied to meet the needs of the crop, the salts are never carried deep in the soil. The high rate of evapouration found in these areas hastens the accumulation of salts in the upper layers of the soil. If uncorrected, the condition may become so severe that only salt-tolerant crops, like sugar beets, can be grown.

Salts from evapourated irrigation water have accumulated to such a high level in this field in the desert biome of California that they stunt the growth of the cotton plants.

The situation can be corrected by using enough addition irrigation water to flush the salts deep into the soil. Unfortunately, many desert soils are shallow and underlain by layers that are relatively impervious to water. Irrigation water that does not evapourate accumulates in the soil, and slowly the water table rises to the surface (if not at the point where the irrigation is applied, then farther down the valley). Soon fields become waterlogged with salty water and, unless steps are taken to drain the water away, productivity quickly declines. Salt tolerance can be engineered into crops that normally are intolerant of saline soils. In 2001, scientists at the University of California Davis campus reported that they had created transgenic tomatoes that grew well in saline soils. The transgene was a highly-expressed sodium/proton antiport pump that sequestered excess sodium in the vacuole of leaf cells. There was no sodium buildup in the fruit.

INTENSIVE IRRIGATION AND WATER LOGGING

Most anti-dam movements in india have emerged from move ments of people facing displacement due to the submergence of large areas upstream of dam sites. These struggles are expressions of a conflict of interests between those who bear the social and ecological costs of dams and those who benefit from them.

However, large dams have diverse and complex ecological impacts. and they often generate environmental costs for those very groups who are supposed to be the beneficiaries. Waterlogging and salini sation are twin problems caused by the wasteful use of water. Waterlogging is caused by the interaction of a large number of factors such as irrigation intensity, soil characteristics, drainage. seepage from reservoirs, distributaries and field channels. Since large-scale irrigation systems are linked to the uniformity of water distribution, which enforces the uniformity of cropping patterns. and uniformity in the landscape, waterlogging becomes inevitable in areas with undulating topography and water retentive soils. In such cases, farmers who are supposed to be the 'beneficiaries' become victims of irrigation projects and irrigation authorities. In areas where irrigation has led to the transformation of productive lands into waterlogged wastelands, conflicts arise between farmers and the state. The 'Mitt) Bachao Andolan' in the Tawacomman area is an example of such conflicts." In the Krishna basin, conflic generated by irrigation projects were highlighted by the farmer agitations in the command area of the Malaprabha project.

The Malaprabha project was completed in 1972-73. With the introduction of canal irrigation, nearly 2,364 hectares of land i the project area has become waterlogged and saline' Before the introduction of perennial irrigation, the undulating semi-arid land in the project area was used for growing water prudent crops like jowar and pulses. Due to a sudden change from rainfed agriculture to intensive canal irrigation, the low lying areas have become waterlogged. The cultivation of water demanding crops like hybrid cotton has aggravated the problem. In addition, seepage from canals has also raised the water level. The Malaprabha project includes a storage dam of 1,068 million cum capacity near Saundatti in Belgaum district which feeds the 138 km Malaprabha right bank, 168 km left bank canal, and the Kolachi right bank canal. The soils in the command area are black cotton soils, which have high water retention capacity and are prone to waterloggin, Intensive irrigation of black cotton soils has been known to be prescription for creating wastelands. While irrigation has bet viewed as a means to improve land productivity, in cases like if Malaprabha command area, it has led to the destruction of productivity.

Further, the shift from rainfed food crops to an irrigated cash crop like cotton was expected to improve the prosperity of farmer However, it led to indebtedness as well as loss of fertile fan. through waterlogging.

To utilise the irrigation waters, farmers began to cultivate 'Varalaxmi' cotton which was initially sold at ₹. 1,000 per quintal. Farmers took loans from banks to develop land, purchase seeds, chemical fertilisers and pesticides. The total loan taken by the farmers increased from ₹. 50 lakhs in 1974 to over ₹. 5.5 crores by 1980. The prices of chemical fertilisers increased from ₹. 75 to ₹. 103 per bag. The cost of the Varalaxmi seed increased from ₹. 60 to ₹. 170 per kg. In the meantime the price of cotton crashed from ₹. 1,000 per quintal in 1974 to ₹. 350 per quintal in 1980.

While farmers were caught in the trap of unfulfilled commercial promise, banks demanded repayments of loans, and the irrigation authorities demanded a development tax known as betterment levy of ₹. 500 to ₹. 1,600 per acre. The water tax was raised from ₹. 18 to ₹. 30 per acre for jowar, ₹. 18 to ₹. 50 per acre for Varalaxmi. A tax of ₹. 10 per acre was fixed even if water was not utilised. For the farmers, this amounted to gross injustice, since they had not benefited from the irrigation project. In addition, the compensation for acquiring land for the dam and canals had not been paid to 75 per cent of the farmers even after seven to eight years.

The farmers therefore organised themselves as the Malaprabha Niravari Pradesh Ryota Samvya Samithi' (Co-ordination Committee of Farmers of Malaprabha Ittihsyrf Area) in March 1980. When the local authorities did not pay heed to the farmers' demands, they launched a non-cooperation movement for non-payment of taxes. The authorities responded by refusing to issue the certificates required by the farmers' children in order to in schools and colleges. On 19 June, the farmers went on a hunger strike in front of the Tebsildar's office in Naragund town. On 30 June, 10,()00 farmers collected to support those on hunger strike. On 7 July, a massive rally was organised in Navalgund, and the farmers went on a hunger strike. Seeing that no response was forthcoming from the authorities, the farmers organised a 'bunch' on 21 July. When 5,000 to 6,000 farmers had gathered in Navalgund, their tractors were damaged and the rally was stoned. The protest then took a violent turn. The angry farmers seized the irrigation office department, burnt down one truck and fifteen jeeps. The police in turn opened fire and a young boy, Basappa Shivappa of Algavadi, was killed on the spot.

In Naragund town, the police opened fire at a procession of 10,000 people, shooting one youth. The protesting farmers responded by beating a police officer and a constable to death. The protests rapidly spread to Ghataprabha, Tungabhadra, and other parts of Karnataka. During the protests thousands of farmers were arrested and forty were killed. Finally, the government had to put a moratorium on the collection of water taxes.and the betterment levy. According to a rough estimate, the concessions granted to farmers to end the Malaprabha agitation amounted to ₹. 85 crores.

However, the high costs of irrigation in the Malaprabha project have been forgotten. No lessons have been drawn for planning water projects, and bureaucrats, technocrats and politicians continue-to get carried away by the euphoria for large dams and intensive irrigation projects. The creation of waterlogged wasteland through intensive irrigation is not specific to the Malaprabha command area. Compared to other projects in the Krishna basin, waterlogging, salinity and alkalinity are most serious in the Tungabhadra project. Nearly 1,500 hectares of land is likely to become waterlogged under the left bank canal.

In the right bank canal 6,000 hectares have been affected by waterlogging. Under the right bank high level canal 12,000 hectares have been affected by waterlogging. In all, 19,500 hectares have been destroyed by waterlogging in the Tungabhadra project within an irrigation period of thirty-five years.

In the Bhadra project, of 1,24,392 hectares irrigated, 7,900 hectares have been devastated by waterlogging. In the Malaprabha project, of the total potential of 2,12,086 hectares, only 12,186 hectares have been actually irrigated. Of this irrigated area, 50 per cent has been waterlogged. In the Ghataprabha project, where a higher average has been irrigated than what was actually planned, out of 3,58,542 hectares irrigated, 19,948 hectares have been devastated by waterlogging. In Andhra Pradesh, under the Nagarjunasagar project (NSP) the groundwater level has risen alarmingly within ten years, thereby indicating a trend towards waterlogging. In Maharashtra, the Maharashtra Irrigation Commission claims that 28,000 hectares of land have been affected by waterlogging in the Deccan Canals, *i.e.*, Nira and Mutha Canals. The irrigation commission of 1976 had estimated the total waterlogged area in the basin at 7,828 hectares (6,583 hectares in Karnataka and 1,245 hectares in Maharashtra) and 15,502 hectares affected by salinity. At present 4,45,985 hectares have been affected by salinity. Waterlogging is ecologically linked to large dams because large darns involve the transport of huge quantities of water for intensive irrigation. In fact, the primary rationale given in defence of large dams is to induce a shift from protective irrigation, which is ensured by indigenous irrigation systems, to intensive irrigation for commercial crops. The inevitable ecological impact of overuse of water for irrigation is a build up of water beyond the drainage capacity of the ecosystem. The need for artificial drainage systems arises because the natural drainage processes of the local ecosystem are violated. Waterlogging is thus a symptom of the conflict between water use in the commerciaVmarket economy, and water use for the maintenance of the water cycle including a balance between water entering an ecosystem and water leaving it. By violating the ecological laws of water flow, large dams lead to ecological destruction on the one hand and political conflict on the other.

RIVER DIVERSIONS AND REGIONAL CONFLICTS OVER WATER

Large dams are constructed for allowing major diversions of water from the natural drainage flow of the river. These diversions result in a major change in the distribution patterns of water in a basin, especially when they involve inter-basin transfers. They therefore generate new conflicts over the distribution of water between different regions. Regional conflicts become inter-state conflicts, and are rapidly enmeshed in inter-state and centre-state politics. The Telugu Ganga Canal, which takes off from the Srisailam Dam, is probably the most conflict-ridden river diversion project in contemporary India.

Krishna is the second largest river of peninsular India. Its catchment lies in the Western Ghats and it flows east through the states of Maharashtra, Karnataka and Andhra Pradesh. Krishna is an inter-state river, and conflicts have arisen between the co-riparian states over the allocation of its waters to their respective territories for purposes of development. The Krishna basin like other regions of India had indigenous irrigation works such as tanks, wells and anicuts. There were nearly 27,000 small tanks and diversions on the Krishna river system, mostly in Andhra Pradesh and Karnataka. To these were added new canals during the colonial period for commercial agriculture. These were:

- The Krishna Delta Canals built in 1855.
- The Nira Canals in Maharashtra constructed in I885 irrigating about 150,000 acres.
- The Kurnool Cuddapah Canal in Andhra Pradesh built in 1886, irrigating 100,000 acres.

The majority of the area irrigated by the Krishna Delta Canals and Kurnool Cuddapah Canal (1.11 million acres) lies outside the basin of the Krishna. In 1951, the status of the diversion of Krishna waters was as follows: 411.4 TMCF of water was diverted annually for the irrigation of 2,302,377 acres. Of this 290.1 TMCF was used by Andhra Pradesh, 430 TMCF by Maharashtra, and 78.3 TMCF by Karnataka.

After independence, the large-scale diversion of river waters increased. In July 1951 the Planning Commission convened an inter-state conference to discuss the utilisation of Krishna waters. The dependable annual flow in the Krishna basin based on the recorded gaugings at Vijayawada was agreed at 1,715 TMCF so the balance of flow for new projects remained 970.5 TMCF which was rounded off to 1,000 TMCF and allocations were made between the different states as follows:

For the balance flow in excess of 1,000 TMCF, if any, the allocation for the above states was in the ratio 30:30:1:39. The state of Bombay was allowed to divert the waters to the west across the Western Ghats for the hydro-electric project at Koyna up to a limit of 67.5 TMCF. The agreement provided for a review of the allocations after twenty years. In 1953, states were; reorganised on a linguistic basis, Madras was divided into Andhra and Madras. In 1956 the state of Andhra Pradesh was created by the merger of parts of Hyderabad and Andhra. As a result of territorial changes, the riparian states sharing the Krishna waters are Maharashtra, Karnataka and Andhra Pradesh. An inter-state conference was convened in New Delhi under the auspices of the Union Minister of Irrigation and Power on September 1960, to recast the allocations of Krishna waters made in 1981. However, efforts to reach an agreement among the states proved unsuccessful and widely divergent views were expressed by the different states. ~ three man commission headed by N.D. Gulhati was set up. The commission undertook the first ever attempt 'at a basin-wide survey of the technical implementation relevant to water resources development.'

As observed by Tripathi, the Commission in examining the river flow of both these rivers was greatly hampered by the lack of regular reliable and continuous observations of water discharge at various points in river... The Commission stated that the flow records prior to 1936 were based on formulae different from those followed after 1936. Therefore the Commission stated that it [was] not possible to determine the flow for 86 per cent dependability or for 75 per cent dependability or for any other criterion of dependability. Because of the lack of adequate data of river flow, the Commission could not give positive answers to the terms of reference as regards the availability of water supplies on the river systems. State-wise allocation of Krishna waters was, therefore, not possible owing to the lack of scientifically observed data. The Irrigation Minister decided that adequate river data should be collected over a number of years and analysed continuously. However, tentative allocation was made for ongoing projects. In spite of interim re-allocations, conflicts over Krishna waters continued with each state accusing the other of higher withdrawals from the river than its legitimate share. Maharashtra and Karnataka wanted a tribunal set up under Section 3 of the Interstate Water Disputes Act, 1956.

The Bachawat Tribunal was appointed in 1969 to resolve the Krishna water state conflicts. In 1973 the Bachawat Committee gave its award. The availability of water was assessed at 2,060 TMCF and on the basis of 75 per cent dependability, Andhra Pradesh was allocated 800 TMCF. The award fixed a formula for sharing both during surplus and lean years and was binding on the states until AD 2000.

Mrs. Gandhi the then Prime Minister consulted the co-riparian states to provide drinking water to Madras which had been facing severe shortages. The three states readily agreed to part with S TMCF each. The Chief Minister of Andhra Pradesh later hailed the Krishna water supply scheme to Madras as the Telugu Ganga. The agreement was reached on 14 April 1976. On 17 October 1977, it was agreed that a 330 km long open canal would carry 15 TMCF to Madras.

While the decision to supply drinking water to Madras was agreed by all the states, conflicts arose when Andhra Pradesh decided to use the Telugu Ganga project for irrigation. The ₹. 850 crore project now envisages extension of irrigation to 5.75 lakh acres in three districts of Rayalseema-Kurnool, Cuddapah and Chittoor and one district in the Andhra region-Nellore, in addition to the supply of 15 TMCF of drinking water to Madras. Nearly 42.4 per cent of Kurnool district lies in the Krishna basin. Cuddapah and Chittoor as well as the rest of Kurnool lie in the Pennar basin. Karnataka had questioned the diversion of water outside the basin to the KWDT arguing that only in-basin needs should be considered in determining a state's equitable share, a state should be permitted to divert its share of water outside the basin. Andhra Pradesh maintained that out of basin needs are a relevant factor and that diversions outside the basin for irrigation needs only should be permitted. Using precedence from the American Law, the Bachawat Tribunal held that the diversion of Krishna water outside the basin was legal. The river basin as an integral unit was thus substituted by the state as an administrative unit. The conflicting demands and distributive patterns emerging from the integrity of the basin versus the integrity of the state ifs a major reason for inter-state conflicts between riparian states not getting fully resolved.

Another fundamental reason for the intractable nature of river conflicts arises from the rights established through the priority of Project use in time and the rights based on the priority of need in the long term. Andhra Pradesh contends that the diversion of an additional 275 TMCF of Krishna waters to feed the districts of Kurnool and Cuddapah in Rayalseema for irrigation of 2.75 lakh acres, is within the scope of the Bachawat award as the Tribunal permitted Andhra Pradesh to take advantage of the surplus flows down the river at Vijayawada. As the Bachawat Tribunal stated, 'the state of An&a Pradesh will be at liberty to use in any year the remaining water that may be flowing in the Krishna river'.

Karnataka has objected to the Telugu Ganga irrigation scheme on the grounds that its own projects to harness Krishna waters are still incomplete and what appears to be excess, currently, will be used in the future. Karnataka has made it clear that surplus Krishna waters would not be available for the Telugu Ganga project. Maharashtra has also opposed the Telugu Ganga project on the ground that it violates the inter-state agreement reached in October 1977. The government of Maharashtra has observed that the state has vast chronic drought affected areas. Almost 75 per cent of the Krishna basin area in Maharashtra is drought prone and the state has plans to use the Krishna water allocated to it by the Krishna Tribunal. It has, therefore, to make sure that at the time of review of the award, its legitimate claim to the surplus available water in the Krishna river is not in any way jeopardised by pre-emptive efforts to commit this surplus water to projects like the Telugu Ganga. Karnataka- and Maharashtra governments are resisting the project on the grounds that Andhra Pradesh has already used its allocation and the Telugu Ganga project would enable Andhra Pradesh to establish its right on larger volumes of water through prior utilisation.

Andhra Pradesh has already invested ₹. 200 crores and has 5,000 labourers working on the construction of the canal. Of the 406 km length of the canal, 190 km pass through the reserved forests of the Nellamali Range, for which central environmental clearance has not been obtained so far. At present the work is confined to reservoirs and canals in the non-forest areas. Water for the project is to be drawn from the Srisailam Dam through the head regulator at Pothireddypadu, which has a total carrying capacity of 11,000 cusecs.

The first 16 km of the canal is shared with the Srisailam right branch canal. The common canals run up to Bankacherla cross regulator where the Srisailam right branch canal and the Telugu Ganga Canal branch off to the right and left, respectively. The Telugu Ganga Canal is in fact the old Srisailam left branch canal extending into the Segileru Valley. Water to be drawn for the Telugu Ganga project is to be stored in four reservoirs at Yellgodu, Brahamasagar, Somashila and Kandaleru. At Mithakanda? near the Bankacherla regulator, a 100 feet high ridge divides the Krishna and the Pennar basins, where the water would be transferred outside the Krishna basin. The canal would pass through Kurnool and Cuddapah districts from where the water would flow into the Pennar river at Chenumukapalli. The flow down the river would be picked up at the Somashila Dam and passed on to the Kandaleru reservoir before it reaches Madras. The construction work continues even though the controversy over the Telugu Ganga project remains unresolved.

Andhra Pradesh derives its legitimacy from two arguments. First, it claims it is using only surplus waters for the project, and the right to surplus waters had been granted to it by the Tribunal. Second, it claims that if there is scarcity, then the arid drought prone regions of Rayalseema should not be asked to sacrifice irrigation waters. Instead, Maharashtra should be asked to stop diverting large volumes of water out of the Krishna basin into the Arabian Ocean for power generation for industrial centres. The river Krishna emerges in the Western Ghats and flows eastward down the gentle slopes. The western face of the Western Ghats falls steeply down altitudes of 1,000 to 2,000 feet, providing excellent sites for power generation. However, the water used for hydro-electricity has to be diverted out of the basin, and dropped into the sea. Currently, the power projects in Maharashtra which divert water westwards are the Tata and Koyna Hydel Projects. The former diverts 42.6 TMC and the latter diverts 67.5 TMC.

In this conflict between the demands for power generation and the demands for irrigation in drought prone areas, the Krishna Tribunal protected existing diversions while giving priority to irrigation for future use. As it stated: In the Krishna Basin, water is a scarce commodity. Westward diversion of water for power generation seriously restricts the use of water for downstream irrigation.... Power for Bombay and Maharashtra industry is generated at the cost of depriving the low rainfall areas on the eastern side of the water solely needed for irrigation. The Tribunal, however, allowed the expansion of hydel projects on the condition that over a period of twenty years they would return to the existing capacity. When the next Krishna Tribunal meets in the year 2020 to review the sharing and utilisation of Krishna waters, the concepts of justice and rights as related to water will have undergone dramatic changes, as will the basin itself.

DEVELOPMENT OF TENSION IN THE WATER COLUMNS

Convincing evidence that the water in the xylem vessels is often in a state of tension has been obtained by direct observations of vessels under a microscope. The stems of some species of herbaceous plants, especially the Cucurbits, are especially suitable for such observations.

Such a stem of an intact, rapidly transpiring plant can be fastened in position across the stage of a microscope and, by careful dissection, vessels examined individually. If one of the vessels under observation be jabbed with the point of a fine needle, an immediate jerking apart of the water column at the point of the rupture will be seen to occur, indicating that the water in the intact vessel was actually in a state of tension. Interesting evidence that the water in the xylem ducts of woody stemsis, at times at least, under tension has been obtained by means of an instrument known as the dendrograph. This is a self-recording instrument which measures variation in the diameters of tree trunks.

It is so constructed that its sensitivity is very great and that its recordings are not influenced by temperature effects upon the instrument. Dendographs are used principally to measure periodic variations in the diameter growth of trees. However, even in trees in which diameter growth has ceased, slight diurnal periodic variations in the diameter of trees have been found to be of regular occurrence. A record of the periodic variations in the diameter of a tree trunk for aperiod of several days at a season when little diameter growth was occurring. The trunk attained its minimum diameter during the transported through plants, be they the tallest of trees or herbs only a few feet in height. It is also generally agreed that root pressures play a minorrole in this process at least under certain conditions and in some species.There is no doubt that living cells are involved in the phenomena generally described under the term root pressure and there is at least a possibility that living cells of the xylem are in some way essential to the maintenance or operation of the cohesion mechanism.

RATES OF MOVEMENT OF WATER THROUGH PLANTS

The rate which water ascends through the xylem ducts may vary from a movement so slow as to bealmost imperceptible to a speed of at least 75 cm. per minute. By rate is meant the length of the water column which will move past a given point in one minute. In general daily variations in the rate of ascent of water closely parallel daily variations in transpiration rate.

LATERAL MOVEMENT OF WATER

A cell to cell lateral movement of water in a radial direction undoubtedly occurs along the vascular rays in the stems of most species of plants. In woody stems there is probably also a lateral movement of water around the stem in a tangential direction. Except in trees in which the grain of the wood is twisted, the conducting vessels on one side of the tree generally connect with branches on that side of the treeat their upper extremity, and with roots on the same side of the tree at their lower extremity. If no lateral movement of water occurred in woody stems,it would be expected that removal of the roots from one side of a tree would result in a dearth of water, or perhaps even death of the leaves or brancheson that side of the tree. Experiments have been performed on apple, peach, oak, and other woody species in which the roots on one side of the plant were removed in an attempt to determine whether or not lateral movement of the water occurs.

Although the water content and growth of the plants treated in this manner diminished there was no difference in the moisture content of the leaves on the two sides of the tree. Neither did the leaves on the side from which the roots had been removed show any greater tendency to wilt on clear warm days than those on the other side. These results indicate very strongly that lateral movement of water occurs in woody stems, and that the water conductive system of plants acts as a unit system.

10

Crop Nutrition and Soils

Crop residues usually consist of the aboveground part of cereal plants after grain removal. They are potentially rich sources of energy because up to 80 per cent of their dry matter (DM) consists of polysaccharides. Due to the prevalence and intensity of agriculture in most regions of China, crop residues represent a high proportion of total feed for herbivores. It is estimated that 550 million tonne of these resources are available annually in China. However, they are not all well utilized as energy sources at present, since their digestibility is often low. They partly resist rumen microbial action so their digestion is far from complete. Due to their rigid structure and poor palatability, intake of crop residues is low. These constraints are mostly related to their specific cell wall structure and chemical composition, but there are also deficiencies of nutrients essential to ruminal micro-organisms, such as nitrogen, sulphur, phosphorus and cobalt.

NUTRITIVE VALUE OF CROP RESIDUES

Nutritive value is generally determined by feed composition, intake and utilization efficiency of digested matter. Thus, the value of a feed depends on chemical composition, digestibility, intake and efficiency.

Table contains the nutrient content of some cereal crop residues in China. All of these residues, with the exception of peanut hay, have insufficient CP for efficient rumen fermentation (<9 per cent of DM). Apart from sweet potato vines, all have a high crude fibre (CF) or cell wall content (in terms of neutral detergent fibre (NDF)) and low available energy (*e.g.*, NEL). Such high fibre contents are believed to be negatively correlated with voluntary intake, rate of organic matter fermentation, microbial cell yield per unit organic matter fermented, and propionate: acetate ratio in fermentation end products. Crop residues also have a low mineral content, especially P, and are deficient in vitamins. Therefore, supplementation of crop residues before feeding is necessary, in addition to various treatments. In rice straw, another unwanted constituent is oxalic acid, which can cause rumen disorders. However, oxalic acid can be eliminated by alkali treatment, washing or ensiling.

Various crop residues have their own nutritional values and are used for different animal species. Sweet potato vines and peanut hay are relatively rich in protein, available energy and vitamins, and are mainly fed to pigs in most rural areas. According to a survey conducted by Zhou Meiqing, partially feeding pigs with fibrous feeds, including peanut hay and/or sweet potato vines, fresh, dried or ensiled, is a popular practice in Sichuan province, where close to 100 million swine are marketed each year. In this feeding system, fibrous feeds can meet a large per centage of the nutrient requirements: 40 per cent of digestible energy, 70 per cent of proteins and 80 per cent of minerals and vitamins. Wheat straw and rice straw have high contents of cell walls, and are basically used for feeding ruminants. Millet straw and soybean straw, in contrast, are fairly palatable for herbivores, and are mostly used as feed sources for horses, donkeys, mules and rabbits.

Table. Nutrient Content of Some Crop Residues

Crop residue		Analysis on DM basis							
		DM	NE_L	CP	EE	CF	CW	Ca	P
		(%)	(MJ/kg)	(%)	(%)	(%)	(%)	(%)	(%)
Wheat straw	(Ningxia)	91.6	3.27	3.1	1.3	44.7	73.0	0.28	0.03
Maize stovers	(Jiangsu)	91.8	5.23	6.5	2.7	26.2	70.4	0.43	0.25
Rice straw	(Fujian)	83.3	4.11	3.7	1.6	31.0	64.4	0.11	0.05
Sorghum stovers	(Liaoning)	95.2	4.69	3.9	1.3	35.6	74.8	0.35	0.21
Barley straw	(Xinjiang)	88.4	3.69	5.5	3.2	38.2	80.1	0.06	0.07
Soybean straw	(Jilin)	89.7	3.85	3.6	0.5	52.1	74.0	0.68	0.03
Oat straw	(Hebei)	93.0	4.52	7.0	2.4	28.4	72.3	0.18	0.01
Millet straw	(Heilongjiang)	90.7	4.61	5.0	1.3	35.9	74.8	0.37	0.03
Peanut hay	(Shandong)	90.0	5.70	12.0	2.7	24.6	88.8	0.13	0.01
Sweet potato vine	(Yunan)	91.7	5.53	8.4	2.6	19.8	36.6	1.47	0.48

Table. *In Situ* DM and NDF Digestibility (%; 48 hour) of some Crop Residues

Crop residue	ISDMD	ISNDFD
Wheat straw	47.6	42.2
Rice straw	42.9	40.9
Maize stover	50.6	46.9
Peanut straw	77.2	54.6
Barley straw	44.8	-

Considerable information on DM digestibility of crop residues was obtained from universities and research institutes in China. From the data presented in Table, it is clear that *in situ* DM and NDF digestibilities of wheat straw and rice straw are lower than in other residues.

The relatively high lignin content in those residues is probably responsible, at least to some extent, for the limited cell wall digestibility.

Since the 1950s, many *in vivo* digestibility experiments and feeding trials have been conducted in China to determine nutritive values of crop residues. Some results have been summarized elsewhere. It has been concluded that without treatment or nutrient supplementation, feeding most crop residues can just, or barely, meet maintenance energy requirements.

THE PRINCIPLE AND EFFECTS OF STRAW AMMONIATION

The main component of straw is fibre, including cellulose and hemicellulose that can be digested by ruminants. Some cellulose and hemicellulose are bound to lignin and resistant to microbial attack. The role of ammoniation is to destroy this link, so these fractions are available to the animal. Ammoniation usually increases digestibility by 20 per cent and CP content up to 1-2 times. It can also improve palatability and consumption rate. The total nutritional value can be doubled, reaching 0.4-0.5 feed units for each kilogram of ammoniated straw. In addition, ammoniation reduces mould development, destroys weed seeds (*e.g.*, wild oat, false sorghum, etc.), parasite eggs and bacteria.

AMMONIA SOURCES FOR STRAW AMMONIATION

The sources of ammonia to treat straw include anhydrous ammonia, urea, ammonium bicarbonate and aqueous ammonia.

Anhydrous Ammonia

Anhydrous ammonia means "ammonia without water." Its formula is NH_3, and its N content is 28.3 per cent. The normal dosage is 3 per cent by weight of the straw DM. It is the most economical source of ammonia.

The boiling point of anhydrous ammonia is -33.3°C, its vapour density is 0.59 (that of air is 1) and its liquid density 0.62 (that of water is 1). Gas pressure is 1.1 kg/cm^2 at -17.8°C and 13.9 kg/cm^2 38°C. At normal temperature and pressure, anhydrous ammonia is a gas. Expensive pressure containers are required not only to keep it as a liquid, but also to transport and store it. Anhydrous ammonia is a potentially dangerous and toxic material, and stringent safety precautions need to be observed when using it. Its natural ignition temperature is 651°C. If the ammonia content in the air reaches 20 per cent, an explosion from self-ignition could occur. Attention should be paid to possible ammonia explosions, even though it seldom happens.

Urea

The N content of urea is 46.7 per cent. Its formula is $CO(NH_2)_2$. It is decomposed into ammonia and CO_2 by ureases at ambient temperature. The chemical reaction is:

$$CO(NH_2)_2 + H_2O \xrightarrow{\text{Urease enzymes. Ambient Temperature}} 2NH_3 + CO_2$$

Urea dosage needed to treat straw may vary a lot. The recommended dosage is 4-5 per cent urea on DM basis, taking into consideration the effect of ammoniation and costs. Urea can be transported conveniently at normal temperature and pressure. It is harmless to humans. Treating straw with urea does not need complex equipment and the sealing conditions are not as strict as with anhydrous ammonia. It is known that farmers in Bangladesh ammoniate straw in bamboo baskets lined and covered with leaves of a kind of banana. From the *Yellow Cattle Magazine*, it is known that, in Anhui province, technicians use urea as a source of ammonia to treat straw without cover and get good results, which is beneficial for extension of straw ammoniation to rural areas. At present, urea is a widely used source of ammonia in China. Urea is not as effective as anhydrous ammonia for straw treatment, but it is better than ammonium bicarbonate.

Ammonium Bicarbonate

The nitrogen content of ammonium bicarbonate is 15-17 per cent; its formula is NH_4HCO_3. It can be decomposed into NH_3, CO_2 and H_2O at a suitable temperature (above 60°C). The chemical reaction is:

$$NH_4HCO_3 \xrightarrow{\text{Heating}} NH_3 + CO_2 + H_2O$$

The dosage of ammonium bicarbonate, estimated by its N content, is 14-19 per cent of straw DM. But, according to the experiments carried out in Zhejiang and Shanxi Agricultural Universities, using 8-12 per cent of ammonium bicarbonate can give the same result as with 14-19 per cent. Ammonium bicarbonate is a major product of the fertilizer industry and it is readily available at low price. It has a retail price of ¥ 300/ton (compared to more than ¥ 1000 for urea) and it is easy to use. Since ammonium bicarbonate is an intermediate product of urea breakdown, theoretically, in the right concentration, its effect should be similar to urea. Reports from Zhejiang Agricultural University indicate that in the humid south, straw ammoniated with urea showed more mould spots than with ammonium bicarbonate. It does not decompose completely at low temperature, thus in cold climates the effectiveness of treatment with ammonium bicarbonate is not good. When treating with ammonium bicarbonate in an oven, one day is enough, since the temperature reaches 90°C and it decomposes completely.

Aqueous Ammonia

Aqueous ammonia is a solution of ammonia in water. The concentration is quite variable, but the usual value is 20 per cent. At this concentration, the normal dosage is 12 per cent by weight of straw DM. It is only adapted to areas near to fertilizer factories because its low N content makes transport expensive.

Other Sources

Besides the above sources of ammonia, human and animal urine also can be used to treat straw. However, collection difficulties limit practical applications.

METHODS FOR AMMONIA TREATMENT OF STRAW

At present, the methods for popular ammonia treatment of straw in China include stack, silo or bunker, and oven methods. Each uses a different ammonia source.

Stack Method

The procedure for the stack method is as follows. First, an area is selected with an elevated, dry and even surface. This area is covered with non-toxic polyethylene sheet and the baled or loose straw (chopped or whole) is stacked on it. It is better to bale straw or to chop it into pieces. Especially for hard and thick maize stover, chopping before treatment facilitates feeding, saves plastic and reduces the danger of puncturing the plastic. The moisture content of straw should be adjusted to 20 per cent or more during stacking (anhydrous ammonia requires a low-moisture straw compared with urea and ammonium bicarbonate). The high moisture has a positive effect on straw treatment. However, after treatment, it is difficult to ventilate the straw and it can be easily attacked by mould. A wooden bar, which will be pulled out when ammonia injection starts, is placed before stacking so that ammonia can be injected easily and conveniently. The stack is sealed with non-toxic polyethylene sheet and injected with 3 per cent anhydrous ammonia by weight of straw DM. At the end, the hole left when the pipe is removed is sealed with good quality tape. Operators working with ammonia must be trained and should strictly follow relevant regulations to ensure personal safety. Treatment time varies inversely with temperature. For example, more than four weeks are needed at a temperature of 5-15°C, but only one week at >30°C.

Straw treatment with anhydrous ammonia is simple and efficient, but particularly suited for large-scale use. The method also had been widely applied in developed countries. It requires some expensive equipment. For instance, the cost of an EQ144 ammonia truck tank carrying 5.1 tonne is ¥ 16000; a 200-kg ammonia bottle costs ¥ 2500; a 400-kg bottle costs ¥ 4500; and an 8.82 kW four-wheel tractor costs ¥ 8 000. Furthermore, anhydrous ammonia is not well suited for private farmers because of its dangers. In order to explore the best supply system of ammonia for farmers, ammoniation stations have been built in Hebei Xingtai region, Boxiang county; Baoding region, Dingxing county; and Henan Zhoukou region, Fugou county. These stations have accumulated experiences during the past few years to share with other regions.

ROTATIONS AND CROP SEQUENCE

Crop rotation is an ancient agronomic practice compared to technology such as herbicides. Since rotation significantly increases yield, it is a cornerstone of farm management. The benefits to rotation are numerous, complex and still not fully understood. Studying rotation effects is difficult because long-term experiments are needed to exclude shortterm influences of weather, soil productivity and management changes. Good crop rotation has a beneficial impact on weeds, diseases and insects. When a crop is grown continuously, pest populations adapted to that crop will increase. Rotating crops tends to reduce pest build-up, especially immobile types. For example, rotations with high canola frequency have more cruciferous weeds such as stinkweed, but less grassy weeds such as green foxtail or wild millet (Table).

Table. Effect of Crop Rotation on Stinkweed at Scott, SK (Based on 1981, 1985 and 1990 Plant Counts)

Rotation	Stinkweed Plants/m^2 in Canola Phase
Fallow-canola	190
Fallow-canola-wheat	129
Fallow-canola-barley	27
Fallow-canola-barley-hay	23
Fallow-canola-wheat-barley-hay-hay	50

The impact of crop rotation on weeds is partly due to different herbicides available in canola compared to cereals. Alternate crops in the rotation allow different herbicides to be used and thus can provide better overall weed control in the long term. Rotating crops and herbicides also helps to rotate herbicide groups and thus reduce the build-up of herbicideresistant weeds. The recent introduction of herbicide-tolerant canola systems (Roundup Ready, Liberty Link, Clearfield and Navigator) has improved weed control in canola and provided more options for rotating herbicide group use. Similarly, crop rotation can lower the risk of diseases. Short rotation canola tends to have more problems with blackleg and seedling root rot.

Table. Effect of Rotation on Blackleg

Rotation	Blackleg (% of Plants with Basal Lesions)
Fallow-canola	43.5
Fallow-canola-wheat	9.5
Fallow-canola-barley	9.0
Fallow-canola-barley-hay	8.0
Fallow-canola-wheat-barley-hay-hay	4.0

Proper rotation with cereals will reduce canola disease, provided that volunteer canola and cruciferous weeds are controlled in the cereal break crops. Also, rotation with canola can reduce many cereal diseases such as common root rot, take-all and tan spot. Australian research has documented a significant reduction in take-all disease of wheat following canola, indicating that canola has a "biofumigation" effect. Rotation also affects water use. Oilseed crops like canola and flax may use less water than cereals, and thus improve subsequent crop yields in dry areas. Water use is more efficient when crops with different rooting systems are rotated. In a diverse rotation of cereals, oilseeds, pulses and forages, the different kinds of residue returned to the soil can improve soil fertility over the long term. This fertility increase is due to differences in nutrient cycling, soil aggregation and complex interactions between plants, soil and microbes. Chemicals naturally present in plants and released during growth or decomposition can negatively or positively affect subsequent crops (this phenomenon is termed allelopathy). Carefully plan crop rotation to avoid a serious build-up of disease, insects or hard-to-control weeds. Another factor to consider is volunteer growth from previous crops that may cause serious competition and seeding problems. Herbicide carryover from previous crops may also adversely affect canola. No single crop rotation will suit all circumstances. The choice of which crops to grow, and in what sequence, depends to a large extent on the soil and climatic conditions of a particular farm and also on the grower's management skills. In areas with adequate moisture, rotations that include cereals and broadleaf crops maintain pests at lower levels and produce consistently higher yield. An added benefit is that yields are often more stable because crops differ in their response to stress at various stages. Therefore, stress may damage one crop seriously, but others in the rotation may be less affected. Use the information in Table as a guide in selecting a field for canola.

Table. Guide to Field Selection for Crops

Crop Before Canola	Wait Period	Remarks
Barley, canary seed, spring rye, fall rye, oats, wheat, triticale, winter wheat	None	No diseases in common with canola. Can be grown the year before or after canola. Control stinkweed, cleavers, mustard, volunteer canola and other problem weeds in these crops. Consider potential herbicide residue carryover problems. A low probability of a volunteer problem for one year for canola on spring rye and triticale.
Buckwheat	One year	A high probability of a volunteer buckwheat problem exists for one year. A low probability of a damping-off and root-rot problem exists for one year. Where wild mustard was a problem in the buckwheat, a medium probability of a contamination problem for one year.
Canola	Three years	A high probability of a disease problem for sclerotinia for four years, blackleg for three years, white rust in B. rapa varieties for three years, alternaria for two years; and a low probability for root rot for two years and damping off for one year. If wild mustard was present, a medium probability of a contamination problem for several years. Consider potential herbicide carryover problems. A high probability of a root maggot problem for one year following a B. rapa crop.

Corn	Two years	Consider potential herbicide carryover problems. Soil test to a depth of 60 cm (24") to monitor nutrient levels and avoid over fertilization as corn is usually heavily fertilized.
Flax	One year	A high probability of a volunteer problem for one year. A low probability of a disease problem for root rot for two years and damping off and sclerotinia for one year.
Forage legume (alfalfa, clover)	One year	Consider potential herbicide carryover problems. A low probability of a disease problem for three years for sclerotinia, two years for root rot and one year for damping off. Available nitrogen may approximate summerfallow nitrogen levels if legume breaking is done before June 30. Perennial legumes have a relatively high sulphur requirement and may deplete soil supplies. Soil test to determine sulphur status.
Mustard	Two to	A high probability of a volunteer problem for one
	to three years	year, followed by a medium to low probability for two years. Mustard is a contaminant of canola, therefore, control volunteer mustard in the wait period. If wild mustard was present, a medium probability of a contamination problem for one year. A high probability of a disease problem for sclerotinia for four years, white rust for three years, and a low probability for root rot for two years and damping off for one year.
Potato	One year	Consider potential herbicide carryover problems. A low probability of a disease problem for three years for sclerotinia, two years for root rot and one year for damping off. If wild mustard was a problem in the potatoes, a medium probability of a problem for one year. Soil test to a depth of 60 cm (24") to monitor nutrient levels and avoid over fertilization as potatoes are usually heavily fertilized.
Pulse (pea, bean, lentil)	One year	Consider potential herbicide carryover problems. A low probability of a disease problem for three years for sclerotinia and one year for root rot and damping off.
Sugar beet	Three years	Cyst nematodes infest sugar beets causing root deformity, reducing production and sugar content. Nematodes may lie dormant in soil for several years as cysts (also attacks canola/rapeseed and mustard). Three- to four-year rotations among sugar beets, canola/rapeseed and mustard reduce nematode populations in the soil. A low probability of a disease problem for one year for root rot and damping off.

Sunflower	Three years	A high probability of a volunteer problem for one year. Consider potential herbicide carryover problems. A high probability of a disease problem for sclerotinia for four years, and a low probability for root rot and damping off for one year.

EFFECTS OF PRECEDING CROP ON CANOLA

Growing canola on canola stubble usually results in reduced yield compared to canola on cereal, pulse or flax stubble. Studies by Agriculture and Agri-Food Canada at Melfort and Aylsham, SK in the Moist Black soil zone have shown a large yield reduction in growing canola on canola (Table).

Table. Relative Yield of Crops Grown on Various Stubble Types at Melfort and Aylsham

Stubble	Peas	Flax	Canola	Wheat	Barley
% Yield of Check (Crop on own Stubble = 100%)					
Cereal	125	111	152	98	109
Other oilseed	114	109	177	131	138
Pea	100	142	196	147	152

Alberta Management Insights summarized the effect of stubble type on canola yields using crop insurance records. Three million acres of data are included in this study.

Table. Effect of Stubble Type on Canola Yield

Soil Zone	Stubble Type	1992-98 Avg. Yield % Canola on Canola	Range of Annual Avg.
Black (south-central)	Wheat	107	96-130
	Barley	115	99-127
	Canola	100	-
	Summerfallow	118	99-139
Black - Dark Grey (north-central)	Wheat	106	87-133
	Barley	125	109-158
	Canola	100	-
	Summerfallow	121	97-155
Dark Grey - Grey (north-central)	Wheat	138	94-172
	Barley	129	113-150
	Canola	100	-
	Summerfallow	134	112-157
Thin Black	Wheat	109	96-120
	Barley	115	103-131
	Canola	100	-
	Summerfallow	125	107-136
Dark Brown	Wheat	116	96-152
	Barley	119	97-164
	Canola	100	-
	Summerfallow	152	123-219
Dark Grey - Black (Peace region)	Wheat	103	94-109
	Barley	101	91-122
	Canola	100	-
	Summerfallow	110	105-119

In addition, similar data from Manitoba is summarized in Table.

Table. Relative Yield of Major Crops Sown on Selected Stubble Types in Rotation

	Stubble Type					
	Wheat	Barley	Oats	Brassica Napus Canola	Flax	Peas
Relative % Yield (Crop on Own Stubble=100%)						
Wheat	100	109	110	118	114	120
Barley	115	100	110	119	122	122
Oats	114	103	100	124	123	115
Brassica napus canola	114	115	117	100	118	128
Flax	148	148	146	133	100	**

** Insufficient Data

In General

- Canola yield is generally better on cereal stubble than canola. The exception appears to be in the Peace River region, perhaps due to the longevity of established canola diseases there (blackleg and brown girdling root rot) and the prevalence of canola.
- In Alberta, canola on barley appears to be slightly better than on wheat except in the Peace and Dark-Grey zones. Although the increased canola yield on cereal stubble is less than found in scientific experiments, this may be due to limiting factors of weeds, diseases and fertility in some commercial fields.
- In Manitoba, canola on flax and pea stubble had the highest yields.
- Yearly yield variation in the Alberta study is very large, indicating a strong weather influence on rotation response.

ROTATIONAL IMPACTS

Growing canola on canola stubble promotes disease and insect build-up. Light disease or insect levels will usually do little damage in the first crop of canola. However, their build-up in that crop can result in significant damage and yield losses in following canola crops. Yield reductions may occur when canola is grown on other crops susceptible to the same diseases and insects, such as flax, mustard, sweet clover, soybeans, field beans, lentils and sunflowers. As shown in the above tables, canola does well on cereal stubble if volunteer plants are controlled. Also, if cruciferous weeds or volunteer canola were a problem in the preceding crop, disease and insect problems could persist into the canola crop. If the preceding crop is a hay or pasture crop, use an early partial fallow period to facilitate seedbed preparations in the next spring. This will allow some weed control and soil moisture build-up. Late summer timing of sod-breaking may limit subsequent canola production because of depleted soil stored moisture, poor fertility and weed control, and the difficulty in preparing a good seedbed with sod clumps.

Another rotational impact of other crops on canola is the potential for toxins that inhibit the growth and yield of subsequent crops (allelopathy). Research around the world has documented that leachates from wheat, alfalfa and some forage grasses can be toxic to canola seedlings. Recent Australian research has found that wheat varieties can differ in the toxicity of their residues to canola. Wheat residue phytotoxicity may partly explain occasional poor emergence and vigour of canola direct seeded into heavy wheat straw. To reduce residue toxicity, spread wheat straw and chaff evenly behind the combine, and in cases of extreme straw production, bale wheat straw. Growing semidwarf cereal varieties may also reduce straw load.

FACTORS DETERMINING THE CROP COEFFICIENT

The crop coefficient integrates the effect of characteristics that distinguish a typical field crop from the grass reference, which has a constant appearance and a complete ground cover. Consequently, different crops will have different $K_c$coefficients. The changing characteristics of the crop over the growing season also affect the K_c coefficient. Finally, as evaporation is an integrated part of crop evapotranspiration, conditions affecting soil evaporation will also have an effect on K_c.

Crop Type

Due to differences in albedo, crop height, aerodynamic properties, and leaf and stomata properties, the evapotranspiration from full grown, well-watered crops differs from ET_0. The close spacings of plants and taller canopy height and roughness of many full grown agricultural crops cause these crops to have K_c factors that are larger than 1. The K_c factor is often 5–10 per cent higher than the reference (where $K_c = 1.0$), and even 15–20 per cent greater for some tall crops such as maize, sorghum or sugar cane. Crops such as pineapples, that close their stomata during the day, have very small crop coefficients. In most species, however, the stomata open as irradiance increases. In addition to the stomatal response to environment, the position and number of the stomata and the resistance of the cuticula to vapour transfer determine the water loss from the crop. Species with stomata on only the lower side of the leaf and/or large leaf resistances will have relatively smaller K_c values. This is the case for citrus and most deciduous fruit trees. Transpiration control and spacing of the trees, providing only 70 per cent ground cover for mature trees, may cause the K_c of those trees, if cultivated without a ground cover crop, to be smaller than one.

Climate

The K_c values of Table are typical values expected for average K_c under a standard climatic condition, which is defined as a sub-humid climate with average daytime minimum relative humidity (RH_{min}) ≈ 45 per cent and having calm to moderate wind speeds averaging 2 m/s.

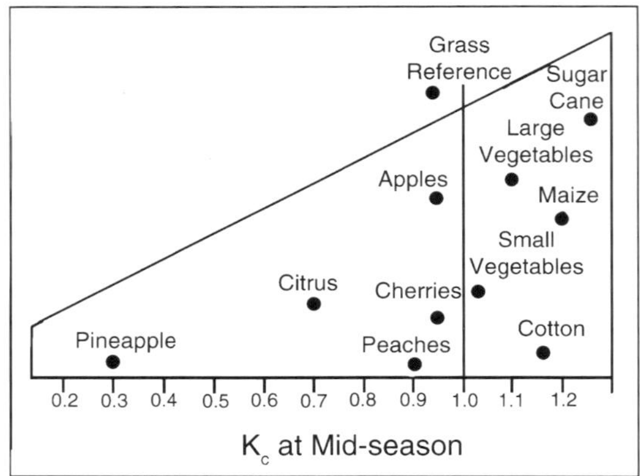

Fig. Typical K_c for Different Types of Full Grown Crops

Variations in wind alter the aerodynamic resistance of the crops and hence their crop coefficients, especially for those crops that are substantially taller than the hypothetical grass reference. The effect of the difference in aerodynamic properties between the grass reference surface and agricultural crops is not only crop specific. It also varies with the climatic conditions and crop height. Because aerodynamic properties are greater for many agricultural crops as compared to the grass reference, the ratio of ET_c to ET_o (*i.e.*, K_c) for many crops increases as wind speed increases and as relative humidity decreases. More arid climates and conditions of greater wind speed will have higher values for K_c. More humid climates and conditions of lower wind speed will have lower values for K_c.

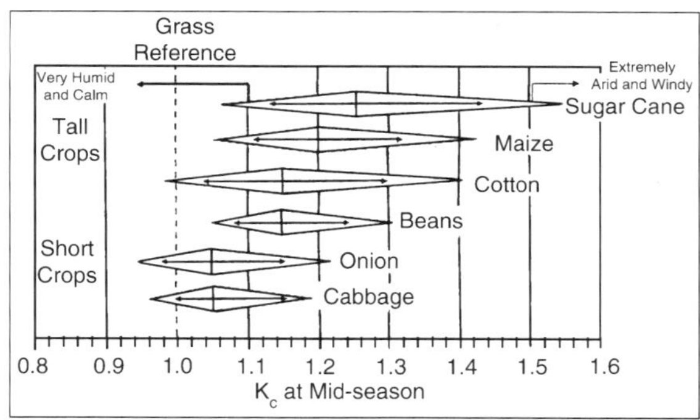

Fig. Extreme Ranges Expected in K_c for Full Grown Crops as Climate and Weather Change

The relative impact of the climate on K_c for full grown crops is illustrated in Figure. The upper bounds represent extremely arid and windy conditions, while the lower bounds are valid under very humid and calm weather conditions. The ranges expected in K_c as climate and weather conditions change are quite small for short crops but are large for tall crops. Under humid and calm wind conditions, K_c becomes less dependent on the differences between the aerodynamic components of ET_c and ET_o and the K_c values for 'full-cover' agricultural crops do not exceed 1.0 by more than about 0.05. This is because full-cover agricultural crops and the reference crop of clipped grass both provide for nearly maximum absorption of shortwave radiation, which is the primary energy source for evaporation under humid and calm conditions. Generally, the albedos, a, are similar over a wide range of full-cover agricultural crops, including the reference crop. Because the vapour pressure deficit $(e_s–e_a)$ is small under humid conditions, differences in ET caused by differences in aerodynamic resistance, r_a, between the agricultural crop and the reference crop are also small, especially with low to moderate wind speed.

Under arid conditions, the effect of differences in r_a between the agricultural crop and the grass reference crop on ET_c become more pronounced because the $(e_s–e_a)$ term may be relatively large. The larger magnitudes of $(e_s–e_a)$ amplify differences in the aerodynamic term in the numerator of the Penman-Monteith equation for both the crop and the reference crop. Hence, K_c will be larger under arid conditions when the agricultural crop has a leaf area and roughness height that are greater than that of the grass reference.

Because the $1/r_a$ term in the numerator of the Penman-Monteith equation is multiplied by the vapour pressure deficit $(e_s–e_a)$, the ET from tall crops increases proportionately more relative to ET_o than does ET from short crops when relative humidity is low. The K_c for tall crops, such as those 2-3 m in height, can be as much as 30 per cent higher in a windy, arid climate as compared with a calm, humid climate. The increase in K_c is due to the influence of the larger aerodynamic roughness of the tall crop relative to grass on the transport of water vapour from the surface.

Soil Evaporation

Differences in soil evaporation and crop transpiration between field crops and the reference surface are integrated within the crop coefficient. The K_c coefficient for full-cover crops primarily reflects differences in transpiration as the contribution of soil evaporation is relatively small. After rainfall or irrigation, the effect of evaporation is predominant when the crop is small and scarcely shades the ground.

For such low-cover conditions, the K_c coefficient is determined largely by the frequency with which the soil surface is wetted. Where the soil is wet for most of the time from irrigation or rain, the evaporation from the soil surface will be considerable and K_c may exceed 1. On the other hand, where the soil surface is dry, evaporation is restricted and K_c will be small and might even drop to as low as 0.1. Differences in soil evaporation between the field crop and the reference surface can be forecast more precisely by using a dual crop coefficient.

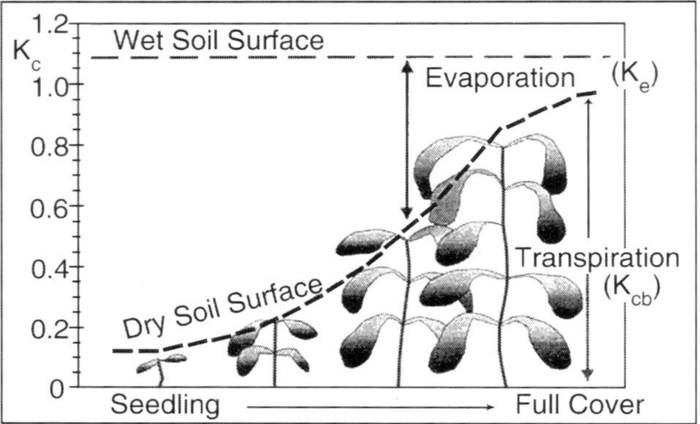

Fig. The Effect of Evaporation on K_c. The Horizontal Line Represents K_c when the Soil Surface is Kept Continuously Wet. The Curved Line Corresponds to K_c when the Soil Surface is Kept Dry but the Crop Receives Sufficient Water to Sustain Full Transpiration.

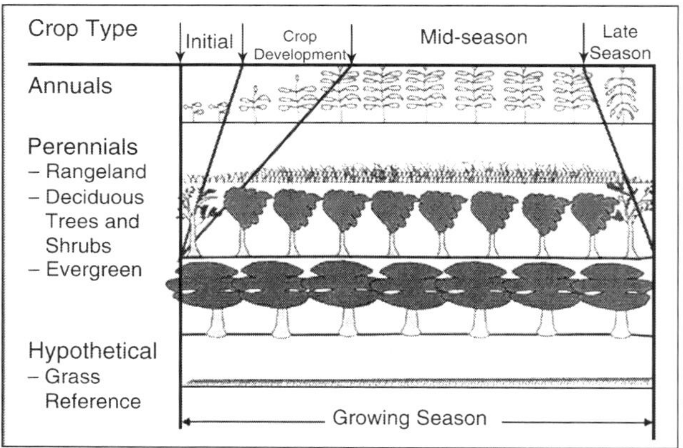

Fig. Crop Growth Stages for Different Types of Crops

CROP GROWTH STAGES

As the crop develops, the ground cover, crop height and the leaf area change. Due to differences in evapotranspiration during the various growth stages, the K_c for a given crop will vary over the growing period.

Initial Stage

The initial stage runs from planting date to approximately 10 per cent ground cover. The length of the initial period is highly dependent on the crop, the crop variety, the planting date and the climate. The end of the initial period is determined as the time when approximately 10 per cent of the ground surface is covered by green vegetation.

For perennial crops, the planting date is replaced by the 'greenup' date, *i.e.*, the time when the initiation of new leaves occurs. During the initial period, the leaf area is small, and evapotranspiration is predominately in the form of soil evaporation. Therefore, the K_c during the initial period ($K_{c\ ini}$) is large when the soil is wet from irrigation and rainfall and is low when the soil surface is dry. The time for the soil surface to dry is determined by the time interval between wetting events, the evaporation power of the atmosphere (ET_0) and the importance of the wetting event. General estimates for $K_{c\ ini}$ as a function of the frequency of wetting and ET_0 are given in Table. The data assume a medium textured soil.

Table. Approximate Values for $K_{c\ ini}$ for Medium Wetting Events (10–40 mm) and a Medium Textured Soil

Wetting Interval	Evaporating Power of the Atmosphere (ET_0) Low 1–3 mm/day	Moderate 3–5 mm/day	High 5–7 mm/day	Very High >7 mm/day
Less than weekly	1.2–0.8	1.1–0.6	1.0–0.4	0.9–0.3
Weekly	0.8	0.6	0.4	0.3
Longer than Once per week	0.7–0.4	0.4–0.2*	0.3–0.2*	0.2*– 0.1*

(*) Note that irrigation intervals may be too large to sustain full transpiration for some young annual crops.

Crop Development Stage

The crop development stage runs from 10 per cent ground cover to effective full cover. Effective full cover for many crops occurs at the initiation of flowering. For row crops where rows commonly interlock leaves such as beans, sugar beets, potatoes and corn, effective cover can be defined as the time when some leaves of plants in adjacent rows begin to intermingle so that soil shading becomes nearly complete, or when plants reach nearly full size if no intermingling occurs. For some crops, especially those taller than 0.5 m, the average fraction of the ground surface covered by vegetation (f_c) at the start of effective full cover is about 0.7–0.8.

Fractions of sunlit and shaded soil and leaves do not change significantly with further growth of the crop beyond $f_c \approx 0.7$ to 0.8. It is understood that the crop or plant can continue to grow in both height and leaf area after the time of effective full cover. Because it is difficult to visually determine when densely sown vegetation such as winter and spring cereals and some grasses reach effective full cover, the more easily detectable stage of heading (flowering) is generally used for these types of crops. For dense grasses, effective full cover may occur at about 0.10–0.15 m height. For thin stands of grass (dry rangeland), grass height may approach 0.3–0.5 m before effective full cover is reached. Densely planted forages such as alfalfa and clover reach effective full cover at about 0.3–0.4 m. Another way to estimate the occurrence of effective full cover is when the leaf area index (LAI) reaches three. LAI is defined as the average total area of leaves (one side) per unit area of ground surface.

As the crop develops and shades more and more of the ground,evaporation becomes more restricted and transpiration gradually becomes the major process. During the crop development stage, the K_c value corresponds to amounts of ground cover and plant development. Typically, if the soil surface is dry, $K_c = 0.5$ corresponds to about 25–40 per cent of the ground surface covered by vegetation due to the effects of shading and due to microscale transport of sensible heat from the soil into the vegetation.

A $K_c = 0.7$ often corresponds to about 40–60 per cent ground cover. These values will vary, depending on the crop, frequency of wetting and whether the crop uses more water than the reference crop at full ground cover (*e.g.*, depending on its canopy architecture and crop height relative to clipped grass).

MID-SEASON STAGE

The mid-season stage runs from effective full cover to the start of maturity. The start of maturity is often indicated by the beginning of the ageing, yellowing or senescence of leaves, leaf drop, or the browning of fruit to the degree that the crop evapotranspiration is reduced relative to the reference ET_o.

The mid-season stage is the longest stage for perennials and for many annuals, but it may be relatively short for vegetable crops that are harvested fresh for their green vegetation. At the mid-season stage the K_c reaches its maximum value.

The value for K_c ($K_{c\ mid}$) is relatively constant for most growing and cultural conditions. Deviation of the $K_{c\ mid}$ from the reference value '1' is primarily due to differences in crop height and resistance between the grass reference surface and the agricultural crop and weather conditions.

LATE SEASON STAGE

The late season stage runs from the start of maturity to harvest or full senescence. The calculation for K_c and ET_c is presumed to end when the crop is harvested, dries out naturally, reaches full senescence, or experiences leaf drop.

For some perennial vegetation in frost free climates, crops may grow year round so that the date of termination may be taken as the same as the date of 'planting'.

The K_c value at the end of the late season stage ($K_{c\ end}$) reflects crop and water management practices. The $K_{c\ end}$ value is high if the crop is frequently irrigated until harvested fresh. If the crop is allowed to senesce and to dry out in the field before harvest, the $K_{c\ end}$ value will be small.

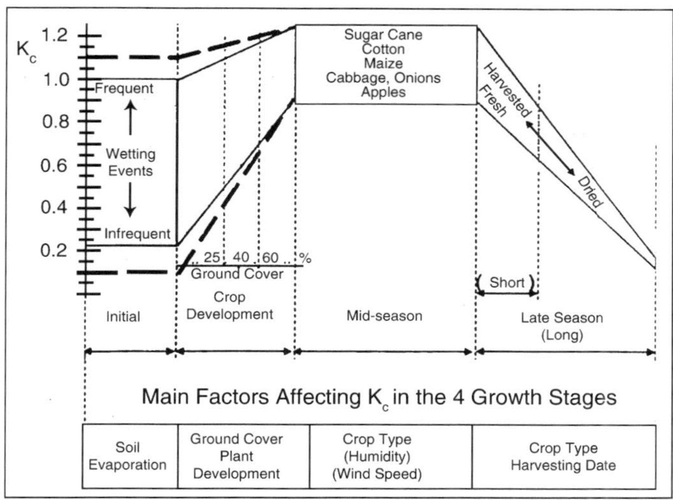

Fig. Typical Ranges Expected in K_c for the Four Growth Stages.

CROP EVAPOTRANSPIRATION (ET_C)

Crop evapotranspiration is calculated by multiplying ET_0 by K_c, a coefficient expressing the difference in evapotranspiration between the cropped and reference grass surface. The difference can be combined into one single coefficient, or it can be split into two factors describing separately the differences in evaporation and transpiration between both surfaces. The selection of the approach depends on the purpose of the calculation, the accuracy required, the climatic data available and the time step with which the calculations are executed. Table presents the general selection criteria.

Table. General Selection Criteria for the Single and Dual Crop Coefficient Approaches

	Single Crop Coefficient Kc	Dual Crop Coefficient Kcb + Ke
Purpose of calculation	• Irrigation planning and design • Irrigation management • Basic irrigation schedules • Real time irrigation scheduling for non-frequent water applications (surface and sprinkler irrigation)	• Research • Real time irrigation scheduling • Irrigation scheduling for high frequency water application (microirrigation and automated sprinkler irrigation) • Supplemental irrigation • Detailed soil and

	hydrologic water	balance studies
Time step	Daily, 10-day, monthly (data and calculation)	Daily (data and calculation)
Solution method	Graphical pocket calculator computer	Computer

SINGLE AND DUAL CROP COEFFICIENT APPROACHES

Single Crop Coefficient Approach (K_c)

In the single crop coefficient approach, the effect of crop transpiration and soil evaporation are combined into a single $K_c$coefficient. The coefficient integrates differences in the soil evaporation and crop transpiration rate between the crop and the grass reference surface. As soil evaporation may fluctuate daily as a result of rainfall or irrigation, the single crop coefficient expresses only the time-averaged (multi-day) effects of crop evapotranspiration. As the single K_c coefficient averages soil evaporation and transpiration, the approach is used to compute ET_c for weekly or longer time periods, although calculations may proceed on a daily time step. The time-averaged single K_c is used for planning studies and irrigation system design where the averaged effects of soil wetting are acceptable and relevant. This is the case for surface irrigation and set sprinkler systems where the time interval between successive irrigation is of several days, often ten days or more. For typical irrigation management, the time-averaged single K_c is valid.

DUAL CROP COEFFICIENT APPROACH (K_{cb} + K_e)

In the dual crop coefficient approach, the effects of crop transpiration and soil evaporation are determined separately. Two coefficients are used: the basal crop coefficient (K_{cb}) to describe plant transpiration, and the soil water evaporation coefficient (K_e) to describe evaporation from the soil surface.

The single K_c coefficient is replaced by:

$$K_c = K_{cb} + K_e \quad (57)$$

where,

K_{cb} basal crop coefficient,

K_e soil water evaporation coefficient.

The basal crop coefficient, K_{cb}, is defined as the ratio of ET_c to ET_o when the soil surface layer is dry but where the average soil water content of the root zone is adequate to sustain full plant transpiration. The K_{cb} represents the baseline potential $K_c$in the absence of the additional effects of soil wetting by irrigation or precipitation. The soil evaporation coefficient, K_e, describes the evaporation component from the soil surface. If the soil is wet following rain or irrigation, K_e may be large.

However, the sum of K_{cb} and K_e can never exceed a maximum value, $K_{c\ max}$, determined by the energy available for evapotranspiration at the soil surface. As the soil surface becomes drier, K_e becomes smaller and falls to zero when no water is left for evaporation. The estimation of K_e requires a daily water balance computation for the calculation of the soil water content remaining in the upper topsoil.

The dual coefficient approach requires more numerical calculations than the procedure using the single time-averaged $K_c$coefficient. The dual procedure is best for real time irrigation scheduling, for soil water balance computations, and for research studies where effects of day-to-day variations in soil surface wetness and the resulting impacts on daily ET_c, the soil water profile, and deep percolation fluxes are important. This is the case for high frequency irrigation with microirrigation systems or lateral move systems such as centre pivots and linear move systems.

SOIL STRUCTURE AND CROP GROWTH

Soil physical properties affect root and shoot growth directly and indirectly, the latter for example through poor drainage causing pores to fill with water and plants to suffer from anaerobiosis. Root growth has been described under various soil physical conditions, but relationships have only rarely been established between features such as crop yield, root growth and soil pore size distribution or conductivity, a more aggregate measure.

The difficulty in establishing simple relationships does not mean that soil has little influence upon root growth; rather it points to the complexity of the interactions (*see fig. 2.4 on page 36*) and the internal homeostasis which plants maintain. Roots both elongate and proliferate and spread laterally as they grow and age. Concomitantly some roots die and others become suberised and function as conduits, but not as absorbers, of water and nutrients. Roots can elongate downwards as fast as 8 cm/d, as for example, soybean growing in a silt loam in a rhizotron. Deep-rootedness and maximum rooting depth reflect soil properties (for example, roots will not grow through pores that they cannot deform to a larger diameter than the root).

However, relationships are not often reported. Maximum rooting depth varies with species and soil type. For example, wheat roots penetrated to 0.8 m in heavy-textured soils and to 1.2 m in a loamy sand but it is often found that a variety will have a consistent rooting depth across similar soil types in a particular year or in one soil across several years. Angus *et al.* found that rice and six dryland crops (mung bean, cowpea, soybean, groundnut, Maize and sorghum) extracted different amounts of stored soil water (ranging from 100 mm for rice to 250 mm for groundnut) and that extraction was, in part, related to rooting depth.

The spread of roots with age can be related to the growth (increase in weight) of the whole plant, and to accumulated temperature or growing day-degrees (GDD) (*see fig. 2.5 on next page*); indeed, there is some evidence that temperature influences the direction of newly-appeared roots as well as the rate of appearance and extent of growth. Clearly, however, there are factors other than plant size, temperature and soil which influence root proliferation. Otherwise the plants sown at three different times of year in the same soil in Figure would align their root growth along a single growth-GDD relationship. These other factors, of which day length is probably particularly important, tend to mask underlying relationships between growth and soil structure.

Figure shows that tillage can affect root length, though in this case its effects took three years to develop. (*see on page 37*) Measurable differences in soil porosity developed under two tillage treatments: in the first year both root growth and water infiltration (K approximately 5 mm/h) were the same under minimum tillage and conventional tillage.

By the third year, when differences were measured between roots, infiltration rates were 84 mm/h in minimum tillage and 0.2 mm/h under conventional tillage. Despite the differences in root growth there were no substantial differences in grain yield, reflecting the overall constraint of climate in the semi-arid environment.

One of the clearer associations between soil porosity and plant growth is described by Tisdall and Cockroft and Tisdall. Soil was ameliorated by deep ploughing, inserting gypsum at depth, and incorporation of straw and green manure in the topsoil. This treatment gave 10-fold increases in the number of earthworms and a 4-fold increase in soil pores. Infiltration rates also increased 10-fold relative to untreated soil. Root growth was not measured, but fruit yields from peach trees increased from 18 to 75 t/ha.

Lateral spread of roots in three sowings of 1987, 1988 and 1989; the ratio r_L/R_L (mean of minimum and conventional tillage) increased with Days After Sowing (DAS) according to $r_L/R_L = 1 - 2.33 \times 0.97^t$ (r = 0.96) where t is DAS. are means od both cultivars and tillage treetments for 1st, 2nd and 3rd sowings (O,,, Δ, respectively) of 1987 (open), 1988 (closed and the 1989 sowing (♦).

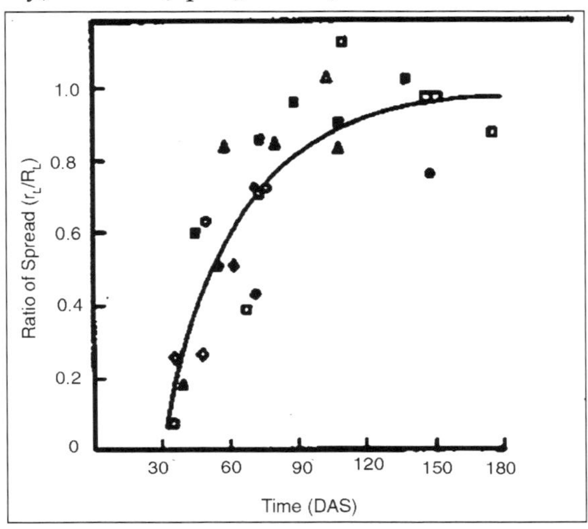

Relationships between root length at each sowing and growing day-degrees (°Cd).

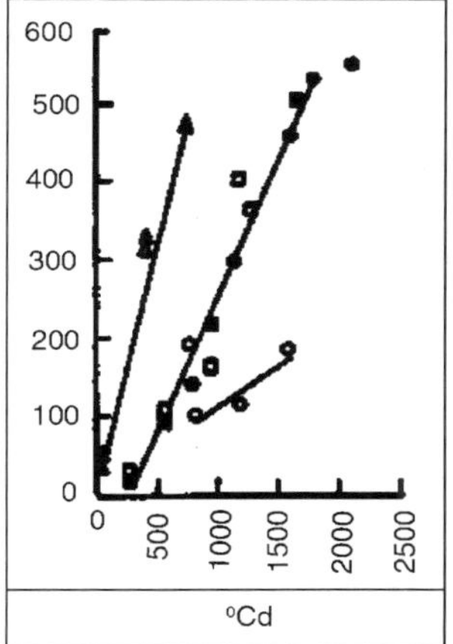

Comparison of total root length per plant in each tillage treetment (minimum tillage/conventional tillage ratio) indicated that root length were lower under minimum tillage than conventional tillage during early growth in 1987, similar in 1988 and higher under minimum than following conventional tillage in tillage in 1989; vertical bars are SE.

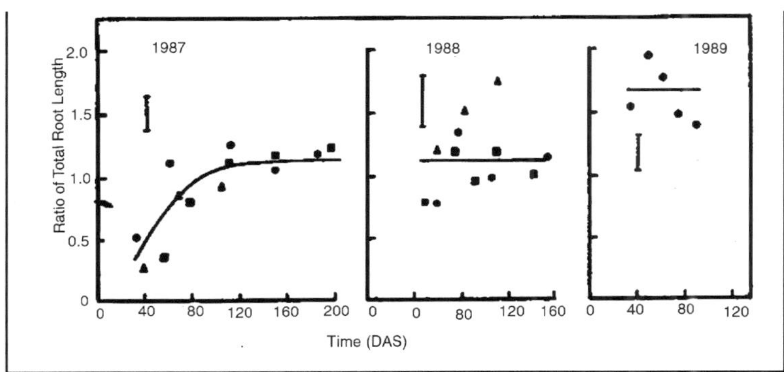

Fig. Spread of Wheat Roots with Age, and Effects on Root Growth of Minimum or Direct Drilling Compared with Conventional Tillage

Table. Values of Air-filled Porosity (Per Cent) and Bulk Density(g cm³) which are Critical and which Limit Root Growth for Various Soils

Texture class	Non-limiting	Critical[1]	Limiting
Air-filled porosity			
Fine loamy	20	10	5
Coarse silty	20	10	5
Fine silty	20	10	5
Clay:			

35-45	15	10	5
>45	15	10	5
Bulk density			
Sandy	1.60	1.69	1.85
Coarse loamy	1.50	1.63	1.80
Fine loamy	1.46	1.67	1.78
Coarse silty	1.43	1.67	1.79
Fine silty	1.34	1.54	1.65
Clayey:			
35-45 per cent	1.40	1.49	1.58
45 per cent	1.30	1.39	1.47

Note: [1] 'Critical' is defined as causing <20 per cent reduction in root growth; 'limiting' is about the value at which root growth ceases.

Increases in soil density or strength retard root penetration and thus limit the volume of soil exploited by the crop and the water available. It is difficult to quantify the relationships between these soil parameters and plant growth. In the cases of bulk density and strength, particularly, a gross measure of either for an undisturbed mass of soil can give only a remote indication of what a root encounters. A determination of gross bulk density does not assess whether a root is growing within a pore (in which case it may deform surrounding soil before its radial environment reaches the density or strength of the gross soil) or if it is growing within the soil material, in which case it has already exerted a radial force equivalent to that measured for the gross soil.

PROPER SOIL SAMPLING EQUIPMENT

Soil sampling to a 60 cm (2') depth can be done with a probe or auger. Do not use flight or screw sampling augers if samples need to be separated by depth since mixing will occur. If testing for micronutrients, ensure the sampling tool is chrome plated or stainless steel and rust-free. Do not bulk samples in metal pails. Clean plastic pails labelled for location and depth work well.

Information sheets, sample bags and shipping boxes are available from soil testing labs and most fertilizer dealers.

PROPER SAMPLING DATE

The most accurate sampling time is just prior to seeding. However, this is practical because time is needed to purchase and perhaps place the fertilizer before seeding. Sampling is commonly done in the spring or fall. Spring sampling is done after the soil has thawed and is no longer saturated from snow melt. Fall sampling can begin once the soil has cooled to 5 to 7Â°C. This helps reduce nutrient content changes due to microbial activity.

The old assumption that N availability does not change over the winter has been proven wrong. Research in Alberta found that available N increased by 56 kg/ha (50 lb/ac) in stubble fields from fall to late winter while the soil was frozen. Summerfallow fields increased by 73 kg/ha (65 lb/ac). This overwinter gain in available N is apparently due to death of soil microbes and subsequent release of available N forms from their ruptured cells. However, the available N gained over the winter was temporary as it was mostly lost in the spring, probably through denitrification. The stubble fields lost 45 kg/ha (40 lb N/ac) in the spring while summerfallow fields lost 73 kg/ha (65 lb/ac). Therefore, the net change from late fall to late spring was minimal.

Late fall sampling tends to more accurately reflect spring NO_3^- (nitrate) contents than early fall sampling, especially for Black soils. Alberta research in the 1980s compared soil samples taken in the fall (early October and early November) to spring samples for 26 stubble fields. Early fall samples averaged 34 kg/ha (30 lb/ac) less nitrate N than spring samples, while late fall samples averaged only about 17 kg/ha (15 lb/ac) less. The early fall samples were also more variable in relation to spring samples. Overall, late fall samples more accurately predicted spring nitrate contents and grain yields than early fall. Spring samples were slightly better than late fall samples for predicting grain yield and N uptake.

In contrast, recent research in Manitoba measured very little change in soil nitrate levels in cereal stubble from early September to freeze-up. North and South Dakota extension soil scientists recommend early fall sampling in view of time constraints, but reduce the N recommendation by 0.2 kg (0.5 lb) for each sampling day prior to September 15.

Another experiment in central Alberta compared the effects of sample timing on phosphorus (P) soil test values. On average over 27 sites, the extractable P increased from 28 kg/ha (25 lb/ac) in early October to 49 kg/ha (44 lb/ac) by early November, and to 50 kg (45 lb) by spring (late April to early May). The relationship between early fall and spring P values was not close and, therefore, it was not possible to simply correct the early fall values. In contrast, research on a Brown soil in Saskatchewan over 24 years found both overwinter increases and decreases in soil P tests, but relatively few were significant. An experiment on irrigated alfalfa on a Dark Brown soil near Lethbridge, AB found significant overwinter increases in organic P. The conflicting results may be related to the differences in soil organic matter content and biological activity, and, therefore, potential for microbial changes to the plant available P pool.

In conclusion, early fall sampling can create higher than necessary fertilizer N and P recommendations due to an underestimation of spring nitrate N and available P, especially in the Black and Gray soil zones. Fertilizer response curves have been calibrated only against spring nutrient contents.

One disadvantage to late fall sampling after soil has cooled to 5 to 7°C is that fall fertilizer banding opportunities become more limited. By the time the samples are taken, dried, sent to the lab, analysed and results returned, the soil may have become frozen or covered with snow. On average over the prairies, soil cools by 1°C every five days in the fall.

CORRECT SAMPLING DEPTH

The appropriate sampling depth depends on the nutrients to be tested. For mobile nutrients such as N and sulphur (S), sampling to the 60 cm (2') depth is usually the most accurate according to research conducted in the 1960s. The fertilizer response database for N was developed with 0 to 60 cm (0 to 24") samples. However, recent research in Saskatchewan and North Dakota indicates that the 0 to 30 cm (0 to 12") depth may be more accurate for N than either 0 to 15 cm (0 to 6") or 60 cm samples. This recent research and the fact that sampling 60 cm is considerably more difficult, supports the 0 to 30 cm depth as a reasonable recommendation for these areas. In contrast, research in Manitoba has documented that the 0 to 15 cm depth is inferior. Separate samples for 0 to 15 cm, 15 to 30 cm and 30 to 60 cm was often recommended in the past, but was rarely done due to additional expense and time.

For immobile nutrients such as P and potassium (K) and most micronutrients, the ideal depth is 0 to 15 cm because the fertilizer response calibrations are based on that depth. If only the 0 to 30 cm depth is sampled, the soil testing labs must use a correction factor to estimate the value for the 0 to 15 cm depth. Since these nutrients are relatively immobile, they tend to remain at the fertilizer application depth. Therefore, the 0 to 30 cm depth may underestimate these nutrients and lead to high fertilizer recommendations.

PROPER SAMPLE HANDLING

Handle samples carefully to prevent accidental mixing and contamination. Mix each sample and spread on separate pieces of clean paper to dry at room temperature. Use a fan if required and avoid additional heat sources. Once dry, fill soil sample cartons or bags with about 0.5 kg (1 lb) of soil, and label with field number and depth.

CONSISTENCY OF SOIL TEST LAB RECOMMENDATIONS

Soil sampling, analysis and interpretation is not an exact science. However, reasonable precision and accuracy is needed in order to make costly fertilizer decisions. Over the years, various growers, agencies or companies have sent duplicate samples to different labs to compare their analysis and recommendations. Unfortunately, widely differing results and recommendations have occurred, causing growers to question the credibility of the labs or the soil testing practice.

Two fundamental reasons contribute to differences in results. First, labs may be using different analytical procedures to measure soil nutrient content. For example, there are several methods to extract soil phosphorus. Or the technique may differ slightly when using a particular method. Over time, labs are harmonizing their test methods and in time this problem may disappear.

Variations in soil test recommendations arise mainly due to the differences in each labs interpretation. The recommended fertilizer amounts at the same soil test level can vary significantly from lab to lab.

This may be due to using:

- Different critical (deficient) soil levels being used
- Regional fertilizer response (calibration) data but modifying recommendations to fit a particular philosophy of fertilizer use or economic payback
- Recommendations from other regions or countries
- A unique system of fertilizer recommendation not based on regional calibration data or economics

Although consistency among labs has been improving since the first comparison studies in the 1980s, further improvements are needed. Calibration data must continue to be collected to account for changes in fertilizer application techniques and changes in other agronomic practices.

Overall, soil testing is a useful agronomic practice. Use labs that base fertilizer recommendations on economics using regional calibration data. Be prepared to question unusual recommendations based on experience and the local knowledge of qualified agronomists. Keep in mind that the accuracy of fertilizer recommendations will always be limited by sampling challenges and the inability to predict the weather of the upcoming growing season.

HOW TO DIAGNOSE NUTRIENT DEFICIENCY SYMPTOMS

In moderate to severe nutrient deficiencies, visible symptoms can indicate the specific nutrient that is lacking. Nutrients that are only slightly limiting often do not show visible symptoms, a situation that has been termed hidden hunger. A systematic diagnosis of visible symptoms is needed to correctly identify limiting nutrients. Symptoms usually appear on either old or young leaves depending on the mobility of the nutrient in question. Chlorosis (loss of green colour, yellowing) and necrosis (death of plant tissue, often leading to white or brown colour) are important visible symptoms. Diagnosis under field situations can be complicated by high field variability, multiple deficiencies, and other causes such as weather, pests and herbicide injury. For example, a sulphur deficiency can easily be confused with Group 2 herbicide injury due to similar symptoms.

PLANT AND TISSUE TESTING

Crop nutritional status also can be assessed by plant and tissue analyses. These methods can supplement, but not replace, soil testing. Plant and tissue analyses measure the nutrient content of above ground plant parts during growth. The values are compared to established ranges for inadequate, adequate and excess levels. Plant tissue testing is suitable for diagnosing crop problems that may be nutritionally related and to identify any nutrients that may be limiting yields. Plant analysis can determine if the fertilizer rate and method of application were adequate to meet crop needs. The disadvantage of tissue sampling is timing after tissue samples are taken and analysed, it may be too late to correct deficiencies in the current crop. No reliable interpretative criteria exist for nutrient ranges in seedling canola. Also, nutrient contents usually differ greatly between different plant parts and ages. Therefore, the proper part must be sampled at the proper growth stage. An adequate sample will contain 50 to 80 plants, depending on the nutrients to be tested and plant part/age. Avoid unusual, dead or stressed plants, as well as those covered with soil or recent sprays. Cut samples with a clean, rust-free knife or scissors. Dry the plant samples on clean paper or plastic at room temperature (do not oven dry). After drying, keep the samples in a paper bag. The following table shows sufficiency levels for most plant nutrients in flowering canola.

Table. Plant Tissue Analysis Interpretative Criteria for Canola (whole above ground plant at flowering)

Nitrogen (N) %	> 2.4
Phosphorus (P) %	> 0.24
Potassium (K) %	> 1.4
Sulphur (S) %	> 0.24
Calcium (Ca) %	> 0.49
Magnesium (Mg) %	> 0.19
Zinc (Zn) ppm	> 14
Copper (Cu) ppm	> 2.6
Iron (Fe) ppm	> 19
Manganese (Mn) ppm	> 14
Boron (B) ppm	> 29
Molybdenum (Mo) ppm	> 0.02

Comparing the plant analysis results from two areas of a field that differ visibly in growth can be difficult to interpret because nutrient content differences can be confounded by growth differences. If the two areas differ mainly in deficiency symptoms, then comparative sampling can be useful. In this case, collect the samples soon after the symptoms appear and before major differences in growth and maturity occur. Plant and tissue analyses need to be interpreted by experienced individuals.

NITROGEN (N)

Nitrogen is the most common limiting nutrient (other than water) for canola production. Therefore, a good understanding of this nutrient is needed to efficiently manage fertilizer N and maximize economic returns.

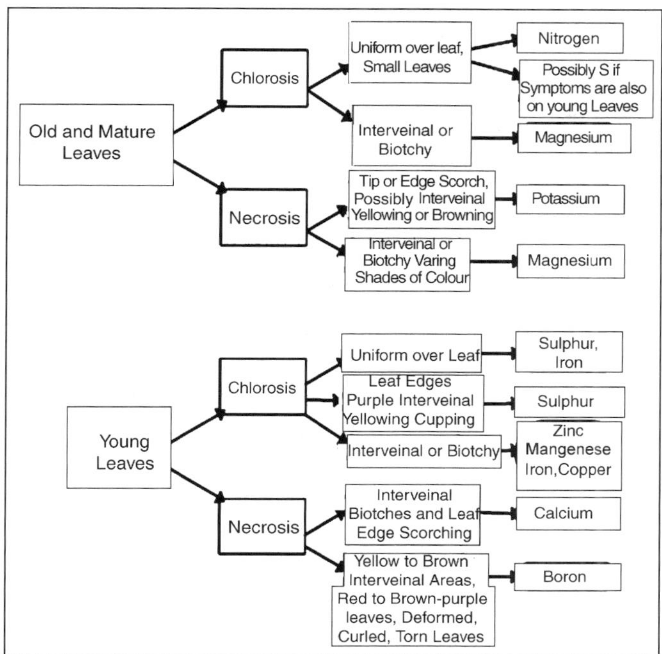

Fig. Diagnosing Nutrient Deficiency Symptoms

SOIL MOISTURE AND MOISTURE USE

Many studies in western Canada and world-wide have reported higher soil moisture under conservation-tillage seeding compared to conventional tillage. The improved soil moisture under conservation tillage is due to reduced evaporation from residue-covered soil as well as increased infiltration rates. Higher soil moisture will often improve germination, emergence, early crop growth, moisture use efficiency and yield. For example, in the 12-year study by Agriculture and Agri-Food Canada at Scott, comparing ZT with CT in two rotations with fallow, canola and wheat, yield increased with ZT where spring soil moisture was increased. In the 36 comparisons over three rotation phases, ZT increased spring soil moisture in nine cases and there were no decreases. Yield was increased in nine cases but decreased in three, and moisture use efficiency increased in six cases with two decreases. The Beaverlodge, AB tillage study found that ZT and RT increased soil moisture in the top 10 cm (4") in dry periods. In excessively wet years, canola growth, moisture use efficiency and yield can suffer under direct seeding.

SOIL TEMPERATURE

Spring soil temperatures are often cooler under conservation-tillage seeding systems than conventional tillage. Most of the western Canadian research studies have found that residue covered soil is 0 to 2°C cooler (daily average temperature) than cultivated soil. In some cases (sunny days), temperatures during midday can vary by 5°C. The colder soil is due to increased soil moisture (water is slower to warm up than air) and more heat reflectance by the residue. Although emergence is sometimes several days longer with direct seeding compared to conventional tillage, the delay usually disappears after canopy closure. Increased soil moisture usually improves yield in spite of the initial cooler temperatures.

In most cases, the cooler soils will not hamper final crop stands or yield. Direct seeders often can seed shallower (which is warmer) due to better moisture, and this largely compensates for temperature differences. However, experience has shown that excessively wet springs can negatively affect canola growth and yield, partly due to temperature effects. Also, frost damage to canola seedlings has been more severe on residue-covered fields when frost followed a warm sunny day.

The greater injury was likely due to lower heat radiation in the critical early morning period under residue-covered soil compared to bare soil. Pay very close attention to achieving uniform residue spreading, preferably with the combine, to reduce cold temperature problems and enable good ground opener performance and seed placement. Research in the Peace River, AB region by Agriculture and Agri-Food Canada found that a narrow strip of bare soil over the seedbed can overcome most of the cold temperature and excessive moisture disadvantage of direct seeding in unfavourable situations.

CANOLA YIELD UNDER CONSERVATION TILLAGE SEEDING SYSTEMS

Research conducted on the prairies has reported variable success with conservation tillage seeded canola-ZT or direct-seeded canola has yielded less, the same or more than conventionally seeded canola.

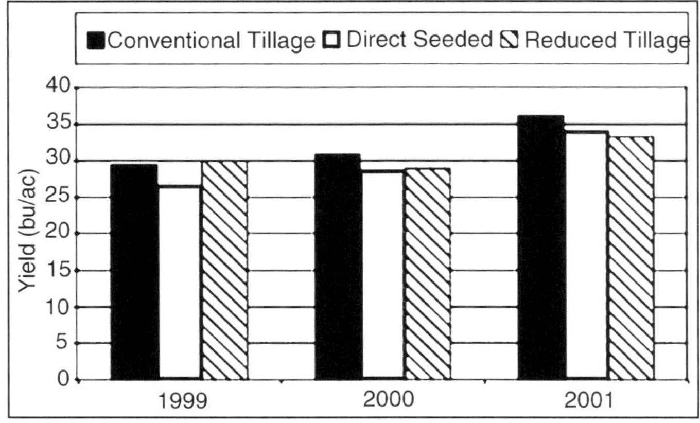

Fig. Canola Yield Comparisons Between Tillage and Direct-Seeded Systems in the Black Soil Zone of Alberta

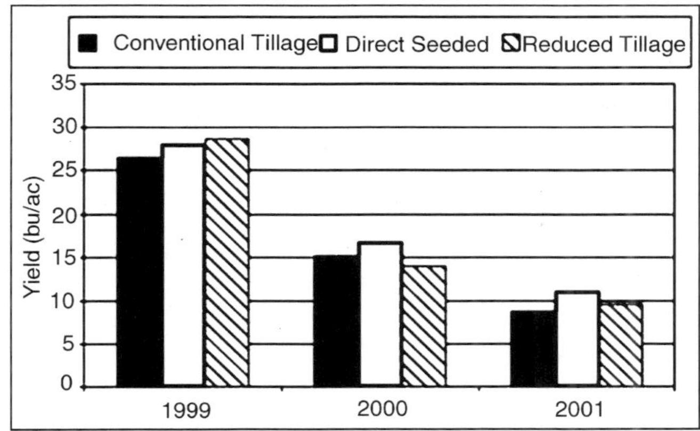

Fig. Canola Yield Comparisons Between Tillage Systems in the Dark Brown Soil Zone of Alberta

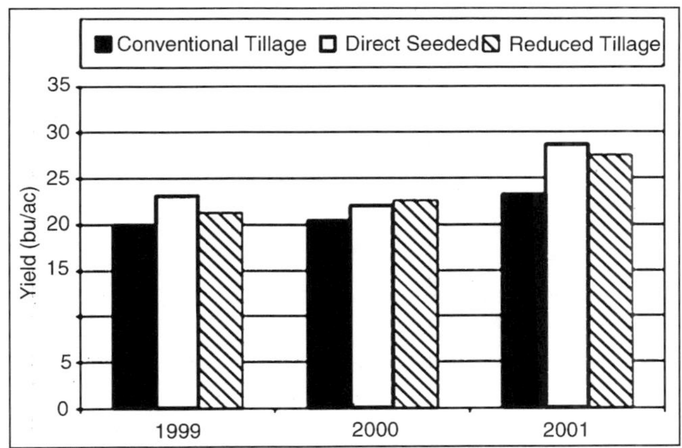

Fig. Canola Yield Comparisons Between Tillage Systems in the Peace River Region - Dark Grey and Grey Soil Zones of Alberta

The many changes in direct-seeding technology and variable weather effects makes it difficult to apply some of the past research findings to the farm level. Perhaps the best indication of canola performance under direct seeding compared to conventional and reduced tillage is from hail and crop agency records. Three years of recent records from the Agriculture Financial Services Corporation in Alberta shows that direct seeding and reduced tillage produced 109 per cent and 108 per cent yield of conventionally seeded canola on average in the major canola growing areas.

Based on 4.6 million insured canola acres seeded on stubble over these three years, 29 per cent, 43 per cent and 28 per cent were conventional tillage, reduced tillage and direct-seeded, respectively. This shows that conservation-tillage seeded systems are more popular now than conventional-tillage systems. Figures illustrate the yield comparisons between tillage systems reported in the major canola growing soil zones in Alberta.

NET RETURNS OF DIFFERENT TILLAGE SYSTEMS

Compared to yield, there have been fewer studies that investigated canola net returns seeded with the various tillage systems. Such studies are difficult since there are numerous combinations of machinery and, therefore, production costs can vary dramatically. Many other confounding variables, other than the tillage seeding system, can also significantly affect the net returns. While direct seeding systems can reduce labour, fuel and some equipment costs, herbicide and other equipment costs may increase. With the widespread adoption of air-drills, the equipment cost has become less of an issue.

Alberta Agriculture Food and Rural Development conducted a survey of 185 growers in 1994 and 1995 to assess the short-term economics of conservation-tillage practices. Based on growers' costs and returns, partial budgets were compared between the systems. Machinery fixed costs were estimated through mathematical formulae.

The main report findings were:

- Zero-and reduced-tillage systems, on average, may have slight economic advantages over conventional-tillage systems.
- Zero-and reduced-tillage systems had marginally better contribution margins, and returns to land, labour and management on average, compared to conventional-tillage systems.
- Contrary to expectations, herbicide costs did not vary consistently between the tillage systems.
- Fuel, repair and depreciation costs increased as the tillage intensity increased. While this is an economic advantage for zero tillers, the more expensive machinery needed for direct seeding tends to offset this.

Overall, conservation tillage systems are suitable for canola production. The higher seedbed moisture can encourage better canola emergence. However, heavy residue fields can create significant problems for good canola seed placement, and increases the risk of frost mortality. Canola seedlings are sensitive to seed placed fertilizer, therefore, ground openers need to be chosen carefully. Further information on conservation tillage practices can be obtained from provincial and federal soil conservation agencies.

Index

A

Absorption 239
Abundance 123
Abundant 2, 16, 78, 80, 145, 152
Accumulated 25
Accumulation 29, 34, 118, 149
Actinomycetes 54, 120, 121, 125, 126, 129, 134, 145, 165, 169
Adequate 148, 149, 150, 153, 158
Adsorption 143
Advent 16, 31
Aeration 28, 31, 130, 133, 148
Azolospirillum 170
Azotobacteria 170

B

Bacterial 202, 203
Basalt 1
Beaverlodge 253
Benecial 77
Biological control of insect 39
Boron 78

C

Canopy 237, 242
Carbohydrates 91
Carbon 8, 64, 65, 66, 67, 68, 69, 70, 203
Carnivores 138
Cemetery 16
Cereals 31, 32
Chlorine 5
Ciat 256
Clostridium 170
Coarse 2, 10
Crumble 75, 187

D

Defaunated 137
Deficient 72, 74, 186, 199
Degradation 2, 30, 119, 120, 122, 123, 145
Denitrification 249
Disintegrate 15
Dissipated 13
Dominant 239
Dung 201, 202, 203, 204

E

Earthworms 20, 97, 98, 99, 102, 103, 104, 105
Elongation 153, 161
Emissions 143
Erosion 9, 28, 30, 32, 54, 61, 122, 151,

200, 201
Exaggerated 131
Examine 15
Excavation 10

F

Farmyard Manure 201
Fertility 28, 30, 31, 32, 216, 217
Fluxes 245
Fragipan 48

G

Gradual 242
Gritty 75, 186
Gypsum 203

H

Haematite 2
Humus 28, 29

I

Igneous 1, 2, 3, 17, 18, 75, 186
Illustrates 17
Immobilization 129, 150
Immobilized 172
Inadmissible 118
Incubation 136
Inoculation 120, 121, 135, 149

J

Jeopardize 33

K

Kaolinite 21

L

Lattice 17
Legumes 31, 32, 33, 34
Leucaena 167
Luvisolic 29, 30, 32

M

Macronutrients 79
Malodorous 127, 129
Manures 47, 170, 181
Micas 2, 3

N

Nematodes 104, 105, 113, 116, 117, 125, 132, 190, 191
Nitric acid 69
Nitrification 103, 117, 125, 164, 166, 167

O

Optimum 120, 122, 123, 130, 132
Origin of Soils 3

P

Paradoxically 78
Parent 49, 50
Polysaccharides 227
Prevail 16
Putrefactive 128, 129, 134

R

Residues 28, 29, 31, 33, 35
Retention 76, 77, 188
Rigid 17, 18
Rigidity 17

S

Scavenger 117
Sedimentary 1, 17, 18
Subsequent 10, 13, 33, 34
Sulphuric Acid 5

T

Tangentially 12, 14
Temperature 61, 70

U

Utilization 42, 180, 205